An Introduction to Economics for Students of Agriculture

SECOND EDITION

Titles of Related Interest

DILLON
The Analysis of Response in Crop and Livestock Production,
2nd edition

BUCKETT
An Introduction to Farm Organisation and Management, 2nd edition

LOCKHART & WISEMAN
An Introduction to Crop Husbandry, 6th edition

WIDDOWSON
Towards Holistic Agriculture

HARBURY
Workbook in Introductory Economics, 4th edition

DOLMAN
Global Planning and Resource Management

EVANS & MILLER
Policy and Practice: essays in memory of Sir John Crawford

Journals of Related Interest

Journal of Stored Products Research

Socio-Economic Planning Sciences

Economic Bulletin for Europe

World Development

Journal of Rural Studies

An Introduction to Economics for Students of Agriculture

SECOND EDITION

by

BERKELEY HILL
Department of Agricultural Economics
Wye College, University of London

PERGAMON PRESS
(A member of Maxwell Macmillan Publishing Corporation)
OXFORD · NEW YORK · BEIJING · FRANKFURT
SÃO PAULO · SYDNEY · TOKYO · TORONTO

U.K.	Pergamon Press plc, Headington Hill Hall, Oxford, OX3 0BW, England
U.S.A.	Pergamon Press, Inc., Maxwell House, Fairview Park, Elmsford, New York 10523, U.S.A.
PEOPLE'S REPUBLIC OF CHINA	Pergamon Press, Room 4037, Qianmen Hotel, Beijing, People's Republic of China
FEDERAL REPUBLIC OF GERMANY	Pergamon Press GmbH, Hammerweg 6, D-6242 Kronberg, Federal Republic of Germany
BRAZIL	Pergamon Editora Ltda, Rua Eça de Queiros, 346, CEP 04011, Paraiso, São Paulo, Brazil
AUSTRALIA	Pergamon Press Australia Pty Ltd., P.O. Box 544, Potts Point, N.S.W. 2011, Australia
JAPAN	Pergamon Press, 5th Floor, Matsuoka Central Building, 1-7-1 Nishishinjuku, Shinjuku-ku, Tokyo 160, Japan
CANADA	Pergamon Press Canada Ltd., Suite No. 271, 253 College Street, Toronto, Ontario, Canada M5T 1R5

First edition 1980

Second edition 1990

Library of Congress Cataloging-in-Publication Data
Hill, Berkeley.
An introduction to economics for students of agriculture/ by Berkeley Hill. — 2nd ed.
p. cm.
Includes bibliographical references.
1. Economics. I. Title.
HB171.5.H63 1990   330'.024631—dc20   89-23239

British Library Cataloguing in Publication Data
Hill, Berkeley.
An introduction to economics for students of agriculture.
2nd ed.   1. Economics. I. Title
330

ISBN 0-08-037497-2 Hardcover
ISBN 0-08-037498-0 Flexicover

Printed in Great Britain by BPCC Wheatons Ltd., Exeter

Contents

Introduction

THIS text aims to provide a simple but effective introduction to economics for students of agriculture and related studies in universities and agricultural colleges. While only a small minority of these students will have studied economics as a subject at school, all are soon made aware that most of the fundamental problems facing agriculture are of an economic nature. A rapid introduction to the discipline is required, yet economics generally forms only part of their total studies so that time for reading is at a premium. A decade of teaching experience without a basic text of suitable complexity and content prompted the writing of this outline of the economic theory of most relevance to this group of students as an attempt at a solution to their particular educational problem.

This introduction to economics aims to describe universal economic principles. It is not a book on the economics of agriculture but one of general economics illustrated primarily by examples drawn from farming and the food industry. Many books concentrating on the economics of agriculture, agricultural policy and farm management assume in the reader some basic knowledge of economics. This introductory text is intended to provide such knowledge and pave the way to the various specialist areas. At the same time, by providing a broad base of theory these specialist studies can be seen in better perspective. Even though a student may be intending to specialise in, say, farm management, his understanding of management problems will be greatly enhanced if he is aware of the arguments put forward for allowing more international competition and trade or of the explanations for inflation and the attempts at its control.

Despite this emphasis on the broad approach to the discipline, students of agriculture will discover that some facets of economics are of greater importance to their studies than others; of particular value are price theory and (because farm management is often a prime interest) production economics. In most introductory texts intended for GCE 'A' level preparation the balance does not favour these areas; instead, attention is switched to a description of the wider economy and its institutions. Intermediate university texts, while usually containing the necessary material, are often found too daunting by the novice. In tailoring an introduction specially for students of agriculture, it has been possible to produce a balance which it is hoped will best serve their needs in terms of content and complexity.

In preparing a second edition I have taken the opportunity of updating the text. Some of the examples needed changing after a decade in which market forces and private enterprise have been allowed to take a greater role in the way that the economy is organised. However, the main change is the addition of a chapter on government policy and agriculture. This does not compromise the general approach of concentrating on the introduction of economic principles, but it does recognise the importance to students of understanding the policy process. Perhaps in no other industry is the impact of government intervention more significant and pervasive. The inclusion of a description of the agricultural policy aims of UK governments and of the European Community and the ways in which these are turned into practical policy programmes was felt to represent a significant improvement to the text. Several reviewers of the first edition were of this opinion.

In any introduction there are always dangers stemming from over-simplification and exclusion. My hope is that serious misconceptions are not propagated here, and that students who find the treatment incomplete will turn to the intermediate texts noted at the end. The general standpoint I have taken to the subject is that of the free enterprise market economy and, although I have included references to centrally-planned economic systems, I am aware that no fundamental criticism of capitalism has been offered. Also perhaps more attention has been shown to Perfect Competition than might be currently fashionable. However, it should be recalled that most of the students for whom this text is intended will find themselves operating in Western economies with agricultural sectors which are capitalist in base, market dominated and where Perfect Competition perhaps comes nearest to reality.

At the end of each chapter is an exercise which makes use of the preceding material, and the completion of these exercises forms an integral part of the teaching function of this text. At the end of the book will be found extended answers to the questions posed in the exercises, a list of essay questions and suggested further reading.

Acknowledgements

COLLEAGUES to whom I owe thanks are Professor D. K. Britton and John Medland, both of Wye College and both of whom have on occasion taught introductory courses jointly with me. I am especially grateful to Alison Burrell for reading the typescript of the first edition and for her many helpful suggestions following hours of discussion. Errors of an economic nature and infelicities in presentation are, of course, fully my responsibility. Past generations of students at the Royal Agricultural College and Wye College would no doubt find much of the text familiar.

CHAPTER 1

What is Economics?

ECONOMICS is concerned with choice at all levels of society; choice by individuals, by firms and by local and central governments. Economics is the study of why choices are necessary and how they are made, a study generally undertaken with the aim of improving in some way the outcome of choices.

At the level of the individual, a person leaving school who is faced with a variety of possible college or university courses has to select just one, knowing that his selection will largely determine how he spends the rest of his life. Having made this major choice, whether he decides on agriculture or economics or law or whatever, he must during his student years allocate his time between the study of his chosen subject and leisure. If he spends more hours playing football, fewer hours are available for work. If his objective is to pass his examinations and yet enjoy some sport, he must choose how he spends his time in order to achieve success in his subject and still have pleasure from his leisure.

Businesses, from family firms to multinational companies, can use their manpower, capital and premises or land in a variety of ways and it is the task of their managements to choose to allocate the firm's resources in the best ways, which usually means in the ways which earn most profit. Taking an agricultural example, a farmer can perhaps grow wheat on his farm and have no other enterprises, or grow grass and keep dairy cows, or maybe he can have a mixture of grass and wheat. It is up to the farmer to choose how to allocate his land and other resources, such as his capital and labour force, between the alternative enterprises, bearing in mind what he is attempting to achieve from his farming. With a farm of a given size, using more land for cereals means that less is available for other crops.

The Government, or State, acting on behalf of society, raises money through taxation or borrowing and spends it on defence, roads, the

1

Health Service and social services, pensions, education and in many other ways. With a budget of a certain size, if the Government wishes to spend more on the Health Service it will have to cut back on something else—say, defence. Just as with individuals, the Government has to choose how to allocate its resources between many possible ways of spending, and by spending money on one thing it loses the opportunity of spending it on another.

The environment is another area where choice by society is involved. For example, society has sometimes to choose between building an airport in an area of natural beauty and having the natural beauty but also suffering congestion at existing airports. By enjoying the convenience of an extra airport, the opportunity of enjoying the unspoilt land is lost.

Opportunity Cost

Whenever choice is involved, taking one alternative automatically means that all the other alternatives have to be left or foregone. If the choice is made to build an airport on a piece of land, the opportunity to use that land for any other purpose is ruled out—for a nature reserve, or for farming, or for a golf course etc.

The cost to society of having an airport may be thought of as what has to be gone without by having the airport. If the land is currently used for farming, being judged the best way to use it before the airport was proposed, then this activity has to be foregone when the airport is built. Costs are commonly thought of in terms of money, but money in this context is only a useful common denominator in which the various items lost and gained by converting farmland into an airport can be expressed and compared. In addition, some of the important costs of having an airport, such as the extra noise involved and the aesthetic loss of unspoiled countryside, are very difficult to express in money terms and may well get left out of the calculation, making nonsense of any precise monetary figure that may be arrived at.

In the example, the land could be used for a whole range of purposes, but only one at a time. If it is used as an airport, the cost of using it as such is really the best alternative use foregone, in this case for farming—this is termed the Opportunity Cost of using the land as an airport.

THE OPPORTUNITY COST OF ANY CHOICE IS THE BEST ALTERNATIVE WHICH IS FOREGONE BY MAKING THAT CHOICE.

Another example is given by the student who can spend an evening drinking, or going to a cinema, or buying a book and studying it. If he chooses the pub but considers that the cinema would be next on his list of enjoyment, so that if he were not drinking he would be watching a film, the opportunity cost of his night in the pub is the visit to the cinema.

Scarcity

Choice by individuals or society becomes necessary because the resources which individuals and society have at their disposal are insufficient to satisfy all their wants simultaneously and completely. As individuals we often come up against money, or rather spending power as a scarce resource, so we have to choose how we use it. Time is another scarce resource.

Some commodities are not economically scarce; we do not normally have to choose how to allocate oxygen in the air—there is sufficient for us to use as much as we want for all purposes. It has no opportunity cost—we can use more of it by lighting a fire without having to breathe less. Such commodities are called *Free Goods*. Air is very common, but there are less common goods which are not scarce in the economic sense because they are plentiful *in relation to the desire or want for them*. An example might be the broken bicycle frames which adorn ditches and lay-bys.

A COMMODITY IS ECONOMICALLY SCARCE WHEN INSUFFICIENT IS AVAILABLE TO SATISFY ALL WANTS COMPLETELY AND THE QUESTION OF ITS ALLOCATION ARISES.

What is an Economic Problem?

An economic problem is said to occur whenever choice between alternatives presents itself. The Government, faced with decisions on how to allocate its revenue between education, defence, housing or social services, is facing an economic problem. So is the farmer allocating his labour force over a range of jobs. So is the fat lady who has to choose between a range of slimming diets on which to spend her money and time. The objective of her choice is clear—at least on first examination— it is to lose weight. What she has to choose is how to achieve it. This leads us on to consider what lies behind any choice—its objective or final goal.

Objectives of Choice

The woman in the above example appears to have an easily identifiable objective, or goal—that of losing weight. However, by questioning more closely, we may find that the taste of her diet is important too; she therefore has two simultaneous goals, weight loss plus a diet which is not unpalatable. A third goal might be cheapness of the diet because lower food costs mean that she can spend more on other goods and services which give her satisfaction. The choice of diet which the woman eventually makes, then, will be the result of her trying to achieve a whole range of goals simultaneously. It is difficult for an outsider to list all the goals and almost impossible to ascribe to each goal some measure of its relative importance. Yet the woman must have all these goals balanced, probably subconsciously, and her final choice will be her way of best achieving her objectives. Her choice will be rational and not just by chance. If presented with the same range of diets at some other time, as long as her objectives had not changed we would expect her to make the same, or a very similar, choice. Whilst it is difficult to specify her objectives, it is clear that she has some. Economists say that the individual is attempting to maximise her *satisfaction* by choosing as she does and make the assumption that the object of any choice is to get the greatest satisfaction which can be attained from the range of available alternatives.

The objectives of society in any choice are equally complex. This applies both to countries whose economies are mainly centrally planned, such as the socialist countries of Eastern Europe, to the countries such as the USA and UK with systems dominated by private enterprise, and to countries which have a mixture of both systems. Some of the objectives which are followed simultaneously and which are common to most countries are:
  —national security
  —a healthy population

—a high level of employment
—a fair distribution of income and wealth
—housing for all
—stable prices
—protection of the environment
—a high level of education
—economic growth
—individual liberty and the individual's control
  of his own environment (democracy).

Some of these objectives conflict and politicans will find themselves "trading-off" one objective against another. For example, to curb inflation it may be necessary to create a pool of unemployment. To prevent atmospheric pollution it may be necessary to control factories more closely and force them to produce their goods in a cleaner and more costly way; this will push up prices and may retard economic growth. Society has, in its choices, to balance all its competing objectives according to their various weights of importance. In a democracy these weights are indicated in the way that people vote at elections. Politicians are put in power who, broadly speaking, reflect the wishes of society.

The objectives of firms are generally easier to define. Their prime objective is to make profit by providing the customer with what he wants, but other objectives might be the minimisation of risk to the business, or to enhance the prestige of the company. In agriculture a farmer might consider profit to be relatively unimportant as a goal once a level has been attained which gives him an adequate living standard. He may then be more interested in doing what gives him pleasure—such as building up a pedigree herd of cows, or spending three days a week at local markets—and will arrange his farm with these non-profit motives in mind.

While objectives are generally difficult to define precisely, it is clear that individuals, society and firms all have them and that their behaviour is directed towards achieving them. Economics is concerned with how these objectives are approached.

We have dealt with scarcity of resources, choice and objectives. They can be combined into a definition of Economics.

ECONOMICS IS THE STUDY OF HOW MEN AND SOCIETY CHOOSE TO ALLOCATE SCARCE RESOURCES BETWEEN ALTERNATIVE USES IN THE PURSUIT OF GIVEN OBJECTIVES.

The Mechanism of Allocating Scarce Resources

In terms of how countries allocate their scarce resources, two chief types of society can be distinguished: (1) the society with a planned economic system, and (2) the society with a free market economic system. Problems of the allocation of the resources which can produce goods and services (i.e. land and natural resources, the labour force and managerial ability, and capital such as machines, buildings, power-stations etc.) occur in both systems. However, the ways in which the choices are made differ.

For the purposes of comparing a thoroughly planned economy with a free market system it is useful to consider each in the most extreme form, i.e. a totally planned system of production and resource allocation with one relying entirely on the market with no central co-ordination. This approach, which uses "models" of each system, is an attempt to capture the essential elements of a real-world situation without the complexities which would be encountered if we were to examine the details of the allocative process in real countries. The use of models permeates economic theory and will be encountered in this text when we consider the behaviour of individual consumers, of firms, of the whole economy and in a range of other contexts. Although models are often criticised as being "unrealistic", by containing the most important elements of a situation and disregarding the rest they may give us a far better insight into real-world situations and the likely outcome of actions or happenings than if we attempted to grapple with the complexities of the real-world situations without their use.

Basically, in any economic system decisions have to be made on three questions: (1) *what* goods and services are produced. For example, should supersonic passenger aircraft be produced, or should the man-power and other resources involved be used to produce, say, more colour television sets? (2) *how much* of each good and service should be produced? (3) *who should get* what is produced?

The country with a centrally planned economic system can answer these questions by a committee, perhaps called the Central Planning Authority, making estimates of how many shoes, coats, television sets are needed by its population and then sending directives to its factories telling them what and how much to produce. Assuming the Central Planning Authority also knows what resources are available in the country, it will probably also tell the factories how they are to produce their products. The nation's resources will thus be allocated in such a way that what needs to be produced *is* produced. Naturally such an economic system can only work where the State owns the vast majority of the land and capital (machines and buildings etc.) and where the inhabitants do what they are instructed to do. Furthermore, the Central Planning Authority will need to have available a vast range of frequently updated information on consumer preferences, the available resources and the technical details of production on which to base its planning decisions.

In the other extreme type of society, that of the unhindered free-market economic system, there is no central planning of what, how much and for whom production takes place. There is thus no centrally planned allocation of the nation's resources and these resources are owned by individuals, not the State as in the centrally planned economy. What is produced, and how much, is determined by consumers going to shops and buying goods. Surpluses of certain goods will soon disappear because, if shopkeepers cannot sell, say, a type of shoe and have to drop its price, they will not order any more from the factories, so production will be cut back. On the other hand, if not enough of a good is being produced, say, not enough colour television sets, so that consumers are waiting for delivery, shopkeepers will badger the factories for supplies. Consumers' wants are rapidly indicated to the producers, and they reallocate their labour force, capital etc. so that what the consumer wants and is able to pay for is produced. Resource allocation hence is a reflection of the wants of consumers; if consumers want a lot of colour television sets, resources will be switched to producing more of them.

No country in the real world is either completely centrally planned or operates a completely unhindered free-market system. However, the problem of resource allocation exists in both types of society and in all intermediate types; only the mechanism of making the choice differs. The study of economics—the study of how men and society choose to allocate scarce resources between alternative uses in the pursuit of given objectives—is equally appropriate in *all* types of society.

The Scientific Approach to Economics

A scientific approach to any subject follows the same basic pattern. The scientist becomes aware, through observation of his subject, of a pattern of circumstances. This might be that, when hydrogen and oxygen mix and are ignited, a liquid—water—is produced. Or it might be that roots of a plant tend to grow downwards even if the plant is turned upside down. Or it might be that consumers buy a greater quantity of petrol when its price is reduced. In an attempt to explain or to account for these observations, the scientist develops a hypothesis, or theory. This is a "positive", or testable, statement about a certain sequence of events with the conditions under which the statement is supposed to apply clearly stated. As well as improving our knowledge of the world about us, a successful theory is useful in that it enables the outcome of various occurrences to be predicted.

A positive statement says what was, is, or will be and its validity can be tested by an examination of the facts. For example, "lowering the price of petrol will increase daily petrol sales" can, or at least might be tested. Normative statements, however, are value judgements. For example, "old age pensions should be doubled" is a normative statement. It may be morally right, but it is untestable. Economics as a science is concerned only with positive statements, although this will not prevent economists from holding opinions, just like any other members of society, on what they consider to be ethically right or wrong.

A scientific theory or hypothesis, then, is a positive statement about a certain set of conditions developed by the economic scientist from his observation of the world about him. For example, the observed relationship between the price of petrol and the quantity sold might be expressed by the following theory: "The price of petrol and the daily quantity sold are related so that, if the price of petrol is reduced, a greater quantity is sold per day, and if the price is increased a smaller quantity is sold, all other things being equal". The last part, "all things being equal" (sometimes given in the Latin form *ceteris paribus*) is very important; it means that other influences on the volume of petrol sold, such as incomes of car owners (as people become richer they tend to spend more on petrol), or the other costs of running a car, do not vary. If they did, it would not be possible to tell whether the change in petrol sales had been the result of the change in petrol price *or* changes in the other influences.

A shorthand way of writing the theory might be

$$D_{petrol} = f(P_{petrol})$$

Demand for petrol is a function of (or depends on) the price of petrol. This is called a "functional relationship".

Once the theory has been formulated it can be tested. This is done by making predictions from the theory and seeing whether these predictions fit the facts. A prediction might be that a fall in the price of petrol will result in an increase in the quantity of petrol sold. If the price falls and the quantity sold *does* increase, the facts support the theory. If the quantity sold does not increase, or decreases, the facts do not support the theory and the theory must be modified in light of any new observations, or scrapped in favour of a theory which explains the facts better.

The process of theory formation and confirmation is shown in Fig. 1.1. If this diagram is understood, the basic scientific approach to any subject will have been grasped.

FIG. 1.1   *Theory Formation and Verification*

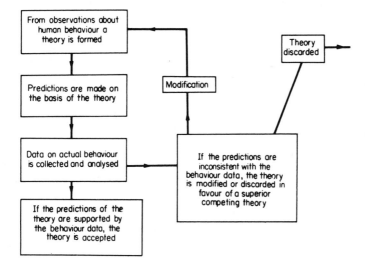

Problems Faced by Scientific Economists

Branches of science differ in the ways in which the predictions from theories can be tested. Two main groups are (a) those sciences where controlled experiments are possible, and (b) those where they are not. Physics, chemistry and most biological subjects fall in the first category, economics and sociology in the second.

If in physics a scientist were attempting to show that as the volume occupied by a gas is reduced its pressure increases, he could in his laboratory control the other influence on pressure, i.e. temperature. If he controlled the temperature of his experiment carefully he would get a consistent result however often he repeated his experiment, although small errors might creep in because of imperfections in his apparatus.

A biologist, wishing to verify the theory that increased nitrogen in the soil causes plants to grow more rapidly, might put a number of plants in a growth cabinet in which temperature, humidity, light etc. could be controlled. Some plants would then be given nitrogen and others, often called the "controls", would not. The scientist would then look for any differences in growth between the plants with and without nitrogen. Not all plants with nitrogen would grow at the same rate—some "biological variation" would occur—but the average, or "mean" growth of plants with nitrogen would be above the average growth of plants without if the theory were valid.

Both the physicist and the biologist in the examples have attempted to eliminate outside influences on their experiments by conducting them in carefully controlled conditions. The economist cannot do this. The basic material of his science is human behaviour, and people cannot be kept in laboratories. If the economist wishes to test his theory about the relationship between the price of petrol and the quantity sold, he cannot control the other factors which determine the quantity sold—such as the incomes of purchasers, or the prices of the other goods on which they could spend their money. Moreover, the economist cannot change the price of petrol—unlike the biologist who could put nitrogen on his plants or the physicist who compressed his gas. The economist has to wait until the price of petrol in the real world changes and then measure the change in quantity sold.* He also has to measure all the other influences on sales and, with the mass of data collected, use complex statistical techniques to analyse the observations and unravel the relationships. Statistics is a highly valuable tool to the economist.

---

* Market surveys *do* partly solve this problem by asking consumers how much they *would* buy if prices were changed; however, they might not actually do what they say they would do.

The lack of laboratory-like facilities for experiments is therefore a major but not insuperable problem faced in the evaluation of economic theories.

A further problem is the variability of human behaviour; this is another, but probably more extreme, version of the biological variation pointed out earlier when the plants-and-nitrogen experiment was discussed. For example, a fall in the price of petrol might induce some people to buy much more, some a little more, some no more, and some less (they might think there was something wrong with it). It is difficult, then, to predict what the response of individuals will be and economic theories are therefore either concerned with the behaviour of groups of people collectively or with the "typical" or "representative" person who takes the form of a group average. This is because, in a group, the odd things that one individual may do are compensated for by the equally odd but opposite things other people may do, an effect sometimes called the *Law of Large Numbers*. Economic theories which apply collectively cannot be disproved by the behaviour of individuals; such theories must be tested for a group or "sample" of adequate size to allow for the odd behaviour of individuals.

Cause and Effect Relationships

Because economics are usually denied laboratory-type experiments it is dangerously easy to draw premature conclusions from data. For example, it can be observed that the number of churches and the number of public houses in towns are positively correlated. A premature conclusion might be that the existence of churches caused people to turn to drink. Further information on town sizes would show that the differences in sizes of the populations could explain both the differences in the numbers of pubs and in the numbers of churches. Incomplete information, therefore, can be misleading and associations do not always mean that cause-and-effect relationships exist.

If a Theory is Shown to be Valid, will it Always be Valid?

The short answer is "no". Economic theories may be shown to hold under a certain set of circumstances, but other circumstances may arise which invalidate them. For example, in the late 19th century a valid theory might have been that the number of cart-horses on farms in England was directly related to the income of the farmer; all other things being equal, farmers with big incomes tended to have many horses. However, with the introduction of the tractor this relationship may have changed, so that the farmers with higher incomes tended to have fewer horses (but more tractors). The theory was not wrong for the earlier period but needed modifying, or superseding by a better theory, when conditions changed.

Conclusion

Economics is the study of how men and society choose to allocate scarce resources between alternative uses in the pursuit of given objectives. Economics is a science which studies human behaviour. It is therefore termed a *social* science, and has to apply a scientific approach without the benefit of controlled experiments. Nevertheless, it is possible to formulate and test theories which help to explain the fundamental economic problem of the allocation of scarce resources.

The first area of economics to be studied contains the theories which try to explain how the individual chooses to allocate his purchasing power (or the scarce resource of the money in his pocket); they are known as Theories of Consumer Choice.

Exercise on Material in Chapter 1

Answer the questions in the spaces provided. If you are unable to answer a question, note the particular difficulty and pass on to the next question. Answers and explanations are given in the Appendix (p. 343).

1.1 "Economics is the study of maximising production." What essentials are missing from this definition?

(i)

(ii)

(iii)

1.2 Briefly describe circumstances where the following commodities are (a) scarce, in the economic sense of the word, and (b) not scarce.

Air	(a)
	(b)
Water	(a)
	(b)
Sand	(a)
	(b)

1.3 Is the following statement normative or positive?

"If the Government raises Income Tax, consumers will have less money to spend."

1.4 What is the difference between a positive and a normative statement?

....
....

1.5 Opportunity cost is ....
....

1.6 A student can either spend an evening studying or playing table-tennis. What is the opportunity cost of—

(a) his evening's table-tennis?

(b) his evening's study?

1.7 A man has the choice of two jobs, with no other alternatives. One pays £1,000/year, but is close to his home; the other pays £1,100/year but will involve daily travel of one hour each way in the firm's bus, which is provided free. What is the opportunity cost of his taking—

(a) the £1,100 job?

(b) the £1,000 job?

1.8 A farmer can plant either wheat or potatoes in a field. The reward for growing either crop is the difference between the receipts from the crop less the costs of growing it. Let this difference be called the "net revenue" for each crop. The farmer decides to grow wheat; what is the opportunity cost of his growing this crop?

1.9 Starting with what you consider to be the first of the following steps, show the usual sequence by arrowed lines.

Testing predictions against observations.

Formation of hypothesis.

Predictions made from hypothesis.

Observations are consistent with predictions.

Hypothesis accepted.

Hypothesis rejected.

Observations not consistent with predictions.

Hypothesis modified or new hypothesis developed.

1.10 (a) You read the following statement in the farming press: "The local NFU Committee feels that farmers on dairy farms of 100-150 acres in Devon who have invested in buildings over the last ten years get lower net incomes now than those who did not." Having in mind the stages shown in 1.9, what process could be followed to show whether this statement was valid?

(b) If the hypothesis is accepted, and present income is found to be inversely related to investment over the last ten years, have you proved that investment in farm buildings causes low incomes?

CHAPTER 2

Explaining the Behaviour of Individuals (Theory of Consumer Choice)

The Concept of Utility

THE economic systems we have been describing exist so that the goods and services we want for living can be provided. In satisfying these wants the goods and services generate satisfaction in people; to use the technical term of economics we say that they generate *utility*. An example of utility is as follows: we have a want for clothes, because we require clothes to keep us warm in cold weather and society also demands that bodies be covered to some extent in public. Articles such as shorts, jeans, etc. have the power to satisfy this want and are thus said to generate utility.

Articles can generate utility without being necessities for living; they need not be "useful" things. Oil paintings give utility because they satisfy the wants of collectors who happen to like them without the paintings being of "use" in any practical sense. Another example is the utility arising from arsenic poison for a would-be murderer. With it the murderer can cause the deaths of many people of value to society. Yet in his manic desire to kill, arsenic gives utility because it satisfies his want, or desire, for a poison. In extreme circumstances the assassination of a malevolent and oppressive dictator may be considered "useful" or "beneficial" to society as a whole, but for the man assassinated the chemical will be anything but beneficial.

Utility is different from demand because demand implies an ability to pay for a commodity. For example, for an impoverished man, a yacht may generate considerable utility in its ability to satisfy his desire for pleasure, but this desire will not be reflected in orders to yacht builders. The poor man will not express any effective demand because he cannot pay for a yacht.

Before proceeding further it must be pointed out that in practice utility cannot be measured in the same precise and universal way that heat can be measured in calories or sound in decibels. If such a workable cardinal unit were available the explanation of human behaviour could well be much easier and public and private choices much more straightforward. However, just because actual utility or satisfaction is difficult or impossible to quantify in absolute units it does not mean that it is useless as a concept for explaining human behaviour. Neither love nor hate can be measured exactly, yet their existence is widely accepted and people's actions are affected by them. For the moment let us assume that the utility generated within a person when a good or service satisfies a want is, at least in theory, measurable. Later we will return to the complications caused by the absence of a practical unit of measurement.

Utility has several important characteristics.

(1). *A commodity generates different utilities for different people, and for the same person at different times.* For example, a pack of hounds gives no utility to a person who is completely against blood sports while the same pack might be a source of great utility to someone keen on hunting.* A pith helmet generates little utility for a native African, but may give considerable utility to a temperate European visitor to Africa. An example of utilities changing with time is provided by thick woollen overcoats, which give little utility in a heat wave, but for their owners generate great utility in the depths of winter. Discussion of utility must therefore take account of the time at which the good is used or consumed.

(2). *For the same person different commodities may satisfy the same want to different extents. They thus generate different amounts of utility and it is possible to rank them in order of utility (or desirability).* For example, a motor car owner may desire a fuel to make his car go. He may be faced with a variety of possible fuels, e.g. good petrol, poor petrol, tractor fuel, household paraffin and vodka. Fig. 2.1 shows that a £1-worth of each of these fuels generates a different amount of utility. Some make it go well, while others hardly allow it to function at all.

---

* To the anti-bloodsport man the pack of hounds could well be a source of dissatisfaction; the pack could be said to generate negative utility or disutility.

Given the choice, the motorist will select the fuel giving the highest utility—good petrol.

(3). *The utility generated by a unit of a commodity varies according to how much a consumer has of that commodity already.* In general, the more of a good a person is already using to satisfy a want (i.e. the more he is already consuming) the less will be the utility generated by additional units of the good. Take the example of clothes. A man may want a dinner suit so that he can go properly dressed to social occasions which require them. He acquires a suit, which gives him considerable satisfaction. However, he knows that occasionally it will be inconvenient to arrange the cleaning of the suit, so he acquires a second one, thereby avoiding this inconvenience and gaining some additional satisfaction.

The two suits satisfy his wants for dinner clothes to a greater extent than one suit, i.e. the two generate more utility than one, but the utility arising from the second suit is much less than that from the first. Fig. 2.2 shows the amount of utility arising from different numbers of dinner suits which this man might possess and the utility coming from the last suit purchased. While the *total* utility increases as the number of suits he possesses increases, it can be seen that the utility generated by each additional suit diminishes. The total utility schedule and the schedule showing the utility coming from the last suit purchased are shown in graphical form by Figs. 2.3 and 2.4.

The Margin

The concept of the *margin* is very important in Economics and is encountered in many facets of the subject. We can illustrate the concept of the margin using the dinner suit example. If the man does not stop at buying two dinner suits but buys a third, that third or additional suit can be regarded as the marginal suit. In more general terms, the last unit acquired is called the *marginal* unit. So instead of saying that the utility derived from additional suits as shown by Fig. 2.4 decreases, we can say that the utility generated by the marginal suit decreases as the number of suits the man possesses increases. The utility coming from this marginal suit is called marginal utility.

This reduction in marginal utility with increases in the number of units already being consumed is so commonly experienced that it has been formulated into a law called the LAW OF DIMINISHING

FIG. 2.1   *Utility from Fuels (hypothetical data)*

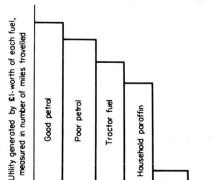

FIG. 2.2   *Number of Suits, Total Utility, and Utility Generated by
Last Suit* (hypothetical data)*

No. of suits	Total utility	Utility generated by last suit
0	0	—
1	10	10
2	15	5
3	17.5	2.5
4	18.75	1.25

* N.B.   Utility generated by the last suit is termed marginal utility.
Utility generated by the nth suit = Total utility n suits - Total utility n -1 suits

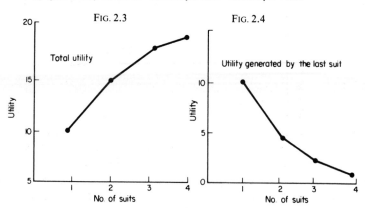

MARGINAL UTILITY. This says that *the utility of additional units of a commodity to any consumer decreases as the quantity of that commodity he is already consuming increases.* Other examples of this law might be; once one possesses a fountain-pen or wrist-watch the satisfaction given by additional pens or wrist-watches is less than that given by the first; if attempting to play tennis, the first tennis ball and perhaps the second tennis ball will generate high utility, but additional tennis balls will give less satisfaction until the stage may be reached where the twenty-fifth or so tennis ball may give no additional satisfaction at all. It may even get in the way and be considered worse than nothing—it could generate negative utility or "disutility".

Examples can be cited where the utility from marginal units first increases and then decreases. This does not necessarily invalidate the Law of Diminishing Marginal Utility as a generality, but calls for a slight modification. If a man has a motor car with no wheels, the marginal utility from wheels will probably increase until he has four wheels to enable him to proceed. However, after this point has been reached, the utility given by each additional wheel may decline. In such cases it is said that marginal utility declines once the "origin" has been attained. In the case of the motor car the "origin" would be when four wheels were possessed (see Fig. 2.5).

FIG. 2.5 *Marginal Utility of a Good Used in Sets e.g. Car Wheels (data hypothetical)*

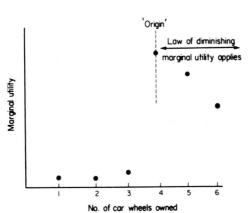

Free Goods

If a consumer is offered an unlimited quantity of an article at no cost, how much of this free good will he help himself to? How many mince pies will a boy eat at Christmas if he is offered a pantryful by his grandmother? The consumer, or the boy, will obviously continue to help himself until the satisfaction given by the last unit (or pie) taken has diminished to nothing. In other words, the quantity of a free good which will be taken is the quantity where marginal utility is zero. Note that to the grandmother the mince pies are unlikely to be free.

Few commodities in life are free goods, that is, they are so plentiful in relation to the wants for them that no allocation is necessary and so they do not have a price, e.g. fresh air and salt water. Most commodities are economically scarce and are therefore priced. The acquisition of them by a consumer is at some cost, and he consequently has to *choose* how to spend his limited funds in the ways to give him most satisfaction. This is, of course, a typical economic problem of allocating a scarce resource (purchasing power) between alternative uses (the range of goods and services open to the consumer) in pursuit of a given objective (to maximise his satisfaction). Economists attempt to explain how consumers do this using Theories of Consumer Choice.

Theories of Consumer Choice

The consumer is of fundamental interest to the economist because it is to satisfy the demands of consumers that production takes place. Paradoxically, the objectives which lie behind the behaviour of consumers and the choices they make are among the most elusive because they are complex and for the most part unmeasurable. In the face of these difficulties economists have developed several theories to explain consumer behaviour, two of which will be described here—firstly utility theory and secondly indifference theory. Both make the reasonable assumption that the objective the consumer has in mind is to get the greatest amount of satisfaction possible from the limited amount of purchasing power he possesses. We begin with utility theory which, while simple in concept, contains some difficulties which the second approach, using indifference curve analysis, overcomes.

A. Utility Theory of Consumer Choice

This theory imagines that it is possible for a consumer to draw up in his own mind schedules of total and marginal utilities for each of the goods which he is able to purchase, although these schedules may be totally subconscious. By subconscious reference to these schedules the consumer can allocate his expenditure in the most pleasurable way.

Let us take a simplified example of a man who can spend his weekly income on only two commodities, bread and cigarettes. For convenience we will assume that one loaf of bread costs 10p and one packet of cigarettes also costs 10p. Hypothetical total and marginal utility schedules for bread and tobacco are shown in Fig. 2.6. In both cases the marginal utility can be seen to decline as the man's consumption of the commodity increases in line with the Law of Diminishing Marginal Utility.

Let us assume that initially the man has 80p to spend and that he buys six loaves of bread and two packets of cigarettes. This will give him a total utility of 450 + 70 = 520 units of utility. Is he spending his 80p in the best possible way? To answer this question we must look at the marginal utility schedules. The sixth loaf of bread which he buys gives 50 units of utility and the second packet of cigarettes generates

FIG. 2.6   *Total and Marginal Utilities of Bread and Cigarettes*
(data hypothetical)

No. units* consumer per week	Bread		Cigarettes	
	Total Utility	Marginal Utility	Total Utility	Marginal Utility
1	100	100	39	39
2	190	90	70	31
3	270	80	92	22
4	340	70	105	13
5	400	60		
6	450	50		
7	490	40		
8	520	30		
9	540	20		
10	550	10		

* The unit of bread is the loaf
  The unit of cigarettes is the packet

31 units of utility. If he were *not* to buy the second packet of cigarettes but instead bought one extra loaf of bread, i.e. a seventh loaf of bread, he would lose 31 units of utility from tobacco but gain 40 units of utility from bread. Thus his total utility would increase by nine units. Seven loaves of bread and one packet of cigarettes is a better choice than six loaves of bread and two packet of cigarettes. Total utility would be 490 + 39 = 529 units.

Should he transfer his expenditure on cigarettes entirely to bread? If he sacrificed his one remaining packet of cigarettes he would lose 39 units of utility and gain from the eighth loaf of bread 30 units. He would thus lose 9 units of utility by transferring the money he would have spent on a first packet of cigarettes to the eighth loaf of bread. In the circumstances, then, the most satisfactory allocation of his expenditure is on seven loaves of bread and one packet of cigarettes.

The example which we have taken is of relatively large units of expenditure, i.e. 10p per unit with a total expenditure of 80p. If we could buy bread by the slice and individual cigarettes, so that the quantity which could be bought would go up in very small steps, we would find that the best allocation of the consumer's money would be achieved when the last penny spent on bread gave the same amount of satisfaction as the last penny spent on cigarettes. Put another way, at the optimum allocation of his spending power the marginal pennyworth of bread will give the same utility as the marginal pennyworth of cigarettes. If a consumer is attempting to maximise the satisfaction he can get from his expenditure, he will always tend to adjust his purchases so that the last penny spent in *any* direction yields the same satisfaction. He will then achieve the greatest amount of satisfaction in total which can be attained with his level of expenditure. This is an example of the *Principle of Equimarginal Returns* which is encountered in many branches of economics, as will become evident from later chapters.

Where goods cannot be easily broken down and bought by the pennyworth, it can be shown (in the Appendix to this section) that a consumer achieves maximum satisfaction from his expenditure when the ratio of the marginal utilities of the last unit of each good is the same as the ratio of their prices. For example, if the price of beer per pint is twice that of a pint of milk, a consumer, in order to achieve maximum satisfaction, will so have to adjust his purchases that the last pint of beer generates twice the utility of the last pint of milk.

Objections to the Utility Theory

The utility theory of consumer choice is open to criticism on the grounds that satisfaction and utility are very difficult to measure. No unit of utility has yet been devised comparable with, say, the calorie as a unit of heat. Calories are definable in an exact way, and apply universally to heat generated by sunlight, oil, combustion, friction etc. In contrast, it is impossible to construct reliable utility schedules for individuals, let alone groups of people, if no practicable unit of measurement exists. Despite the apparent obviousness of the Utility Theory for explaining consumer choice, it is untestable. To meet these objections an indifference theory has been developed, the formulation and testing of which does not necessitate measuring satisfaction or utility in absolute units.

B. An Indifference Theory of Consumer Choice

Again we take a simple model in which a person has the choice of spending his resources on two commodities, in our case beer and milk. Fig. 2.7 shows the various combinations of beer and milk which give the same level of satisfaction to the consumer. Five litres of beer and two litres of milk consumed per week give the consumer the same amount of satisfaction as four litres of beer and three litres of milk or two litres of beer and eight litres of milk. The consumer is indifferent to which combination he consumes. This sort of schedule can actually be constructed (in contrast with the utility schedules given earlier which must remain hypothetical) by giving a consumer five litres of beer and two of milk to start with and then taking beer away from him, bargaining with milk to compensate. Thus in this approach we are not attempting to measure the consumer's level of satisfaction as utility theory implies; we are simply assuming that a consumer knows how much extra of one commodity is required to compensate for the loss of another commodity—in other words, to keep his level of satisfaction the same. This is a more realistic assumption since it does not require the measurement of utility itself. A parallel might be experienced by a River Authority; it is impossible in practice for them to measure the volume of water in their river in gallons, but it is easy for them to watch its level. Gains from heavy rainfall can be compensated for by opening sluices and drought conditions by restricting pumping for irrigation so

FIG. 2.7   *Combinations of Beer and Milk Which Give the Same Level of Satisfaction*

Schedule 1₁

No. litres of beer	No. litres of milk
5	2
4	3
3	5
2	8
1	14

FIG. 2.8   *Combinations of Beer and Milk Which Give the Same Level of Satisfaction*

Schedule 1₂

No. litres of beer	No. litres of milk
5	4
4	5
3	7
2	11
1	20

Note: Although only whole-number combinations are given for simplicity, it is assumed that both commodities are available in fractions of litres so that intermediate combinations involving part litres could be shown in an expanded schedule, i.e. 1.6 litres of beer and 2.5 litres of milk could appear in Schedule 1₁.

FIG. 2.9   *Indifference Curves*

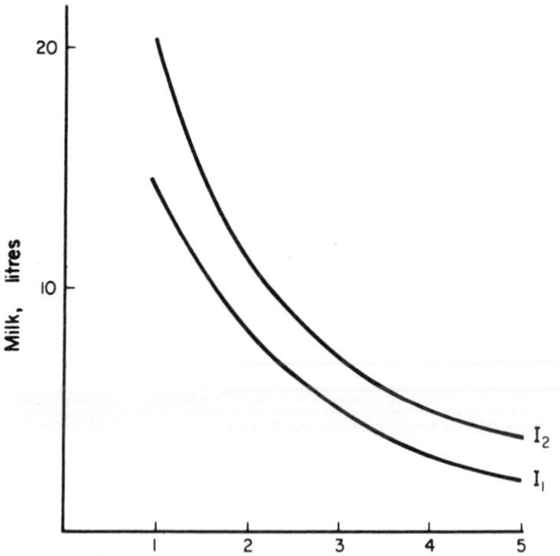

N.B.   An indifference curve shows the combinations of two goods which give the same level of satisfaction

Curves further from the origin correspond to higher levels of satisfaction

that the river level is kept constant. As a consequence, the volume of water in the river is also kept constant although the precise gallonage is not known.

You will notice from Fig. 2.7 that as we reduce the weekly consumption of beer, the consumer demands greater and greater quantities of milk to compensate. For example, when we reduce his consumption of beer from five litres to four, we are only required to give him one extra litre of milk to compensate. However, if we wished to reduce his quantity of beer from two litres to one, he would demand an extra six litres of milk to compensate, i.e. to increase his weekly consumption from eight litres of milk to fourteen. The number of units of one good (milk) required to compensate a consumer for the loss of one unit of the other (beer) i.e. to keep him at the same level of satisfaction, is called the *Marginal Rate of Substitution*. It increases as the consumer's consumption of beer decreases and decreases as his consumption of beer increases. The MRS can also be measured by the number of units of one good (litres of milk) that a consumer is willing to *give up* to gain one unit of the other (litre of beer).*

Another indifference schedule could be constructed by giving the consumer five litres of beer and four of milk to start with. This would mean that he was initially on a higher level of satisfaction than with the previous combinations because the quantity of milk is greater. A schedule such as Fig. 2.8 might result. Obviously any combination in the scheduled Fig. 2.8 is superior to any combination in Fig. 2.7. Plotting these schedules gives us the *indifference curves* we see in Fig. 2.9. Indifference curves further away from the origin imply higher levels of satisfaction.

---

* By convention, the MRS is estimated with respect to changes by one unit in the consumer's holding of the good on the horizontal axis of indifference curve maps. In the example chosen, beer is put on the horizontal axis and the MRS measures the varying numbers of litres of milk required to compensate for the loss (or gain) of one litre of beer. Because of this convention, the MRS of goods at any point on an indifference curve is numerically the same as the slope of the curve. The slope of the indifference curve, and hence the MRS, declines as we move along the horizontal axis away from the origin. To be more mathematically precise, the slope (and MRS) should be said to *increase* as we move away from the origin along the horizontal axis. This is because the slope (and MRS) is a negative quantity and a figure of -5, which might be found near the origin, is mathematically a smaller quantity than a figure of -1, which might be found further from the origin. However, this effect of the negative sign is usually omitted and the MRS is said to decrease with distance from the origin.

Fig. 2.10   *Indifference Curves and Budget Line*

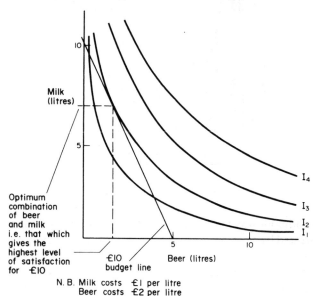

N. B. Milk costs £1 per litre
Beer costs £2 per litre

The following characteristics should be noted:
1. Indifference curves cannot cross;
2. Indifference curves normally have a negative slope;
3. Indifference curves are normally convex to the origin;
4. Assuming that the goods on the axes are readily divisible, an additional indifference curve can always be drawn between two other indifference curves. For example, we could break down the litres and deal with millilitres.

When attempting to explain consumer behaviour using indifference curves, we construct a simple model by assuming that the consumer has a set of them for the two commodities, he has a fixed income—all of which is spent on the two commodities, and that prices are fixed (at least to start with). Fig. 2.10 shows four indifference curves belonging to a consumer for beer and milk.

Fig. 2.11   *Income Effect*

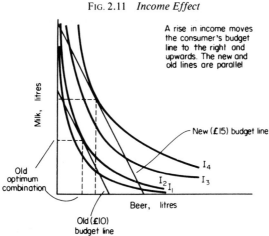

A rise in income moves the consumer's budget line to the right and upwards. The new and old lines are parallel

Milk, litres

New (£15) budget line

I_4

Old optimum combination

I_2 I_1 I_3

Beer, litres

Old (£10) budget line

The optimum way in which the consumer can spend his income can be illustrated by superimposing on to the indifference curve map a line called the *budget line* or *iso-expenditure* line. The budget line shows all the combinations of beer and milk that could be purchased by the consumer at a given level of income and prices. If we assume that the income of the consumer is £10 per week and that milk costs £1 per litre and beer costs £2 per litre,* the budget line can be drawn by connecting the quantities on each axis that can be purchased for £10. Thus, this line connects 10 litres of milk to 5 litres of beer because these are the quantities of goods which could be purchased weekly if *all* his money were spent on one good *or* the other. In addition, any combination of beer and milk lying on this line could be bought by the consumer. It can be seen that at two points the budget line cuts the lowest indifference curve, and the consumer could, if he wished, buy the combinations corresponding to either of these points and thereby experience the level of satisfaction corresponding with indifference curve I_1. The consumer can never buy a combination lying on indifference curves I_3 and I_4 because they lie everywhere to the right of his budget line. The levels

---

* Unrealistic prices have been taken to show that the model for explaining consumer behaviour is not dependent on real-world prices for its effectiveness, although these could have been used.

of satisfaction corresponding to these curves are beyond the reach of his spending power.

However, his budget line *does* touch indifference curve I₂ at one point. At this point the budget line and indifference curve are tangential (i.e. their slopes are the same). The combination of milk and beer corresponding to this point is the optimum (or best) combination the consumer can obtain because it represents the highest level of satisfaction he can attain at the given level of income and prices. There is no point in choosing any combination lying on a lower indifference curve than I₂ because the satisfaction thereby achieved will also be lower. We assume that consumers always aim for the maximum satisfaction available with their resources.* Note that there are many sub-optimal ways for the consumer to spend his income, but only one "best" way.

Income Effect

If the consumer's income is increased to £15 per week, his budget line is shifted to the right and upwards, parallel to the original budget line. This is because, with his higher income, he will be able to buy more beer and/or more milk. This shift in the budget line may enable him to touch indifference curve I₃ and I₄, or some intermediate indifference curve which could be inserted between the existing curves. There is no

---

* At the optimum choice for the consumer the slopes of the indifference curve and budget line are the same. The slope of the indifference curve is the number of units of the good on the vertical axis (y) required to compensate for a change of 1 unit of the good on the horizontal axis (x), and is the same as the Marginal Rate of Substitutions of y for x.

$$\text{Slope of Indifference curve} = \frac{\text{change in no. units of y}}{\text{change in no. units of x}} = \text{MRS} \qquad (i)$$

$$\text{Slope of budget line} = \frac{\text{quantity of y which can be bought for £a}}{\text{quantity of x which can be bought for £a}}$$

Quantity of either good which can be bought is inversely related to its price.

$$\text{Slope of budget line} = \frac{\text{Price per unit of x}}{\text{Price per unit of y}} \qquad (ii)$$

At the optimum choice for the consumer combining (i) and (ii)

$$\text{MRS} = \frac{\text{Price of x}}{\text{Price of y}}$$

guarantee that the ratio between the numbers of litres of beer and milk at the higher optima will be the same as at the optimum established on I_2, since this will depend on the shape of the higher indifference curves. The Income Effect is shown in Fig. 2.11.

Price Effect

The effect of a change in price of either beer or milk will be to change the *slope* of the budget line. If the price of beer halves to £1 per litre, the £10 budget line will now connect 10 litres of beer to 10 litres of milk as the price of a litre of each will now be £1. This has been shown in Fig. 2.12. The new budget line is tangential to I_4 and consequently a new optimum combination of beer and milk is established for the consumer. Note that when beer becomes cheaper and the slope of the budget line changes, the new optimum combination which the consumer can buy with his £10 includes a larger proportion of beer than before. When the price of beer falls, more of it is bought. This is a highly important link between price of a commodity and the amount which consumers are willing to purchase. The relationship between price and quantity which consumers purchase is an important element of demand. If we were to

Fig. 2.12 *Price Effect*

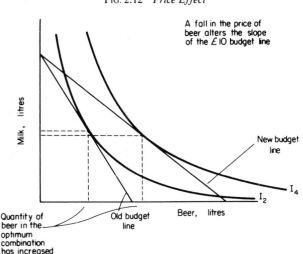

A fall in the price of beer alters the slope of the £ 10 budget line

New budget line

I_4

I_2

Milk, litres

Beer, litres

Old budget line

Quantity of beer in the optimum combination has increased

keep the consumer's income and the price of milk constant by drawing a number of budget lines corresponding with a range of prices of beer, we could construct a table showing the quantities of beer which the rational consumer would buy per week at different prices. Such a table is called a demand schedule. If the price of beer is plotted against the quantity that is bought per week, with price on the vertical axis, the result is a demand curve. This is shown in Fig. 2.13. All other influences on the amount that is bought are assumed to be held constant. The curve shows the amount of a commodity which a rational consumer will purchase, faced with a choice between commodities and changes in the price of the commodity in question. Such curves will be met many times in later sections.

FIG. 2.13   *Amount of Beer Bought at Different Prices of Beer*

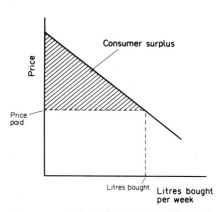

Demand, Marginal Utility and Consumer Surplus

A demand curve shows the quantities of a commodity which a consumer would buy at a range of prices, *ceteris paribus*. But the relationship between price and the amount that consumers are willing to buy can be looked at from another angle, which links back to the earlier consideration of marginal utility. It also shows what the prices would have to be if the consumer were to be persuaded to buy given quantities.

If a rational consumer is offered an additional unit of a commodity, how much will he or she be willing to pay for it? Its purchase will mean diverting spending power from other goods and services, so some opportunity cost is involved. In these circumstances money is a useful common denominator in which the competing ways of spending money can be expressed. Staying with the example of beer, if the consumer is willing to pay a high price for the extra pint we know that the amount of utility which can be derived from the additional pint is great (that is, marginal utility is high). He is willing to forego substantial amounts of other goods and services on which he could otherwise spend this money. Assuming that the Law of Diminishing Returns applies, we would expect the price which a consumer would offer for an additional pint would be smaller if he already has a large quantity. He might be prepared to pay £2 for the first pint, £1.50 for the second, and £1 for the third (assuming he bought them all together and that his rationality did not change as a result of buying and consuming the beer). Thus a demand curve can be interpreted as representing the diminishing marginal utility to a consumer as the amount already possessed increases, with the price standing as a proxy for the marginal utility.

If in practice a consumer can buy all his beer at the same price (£1 per pint) he will buy three pints. The marginal utility of a fourth pint is less than the utility to be derived from spending the money in other ways, so he does not purchase it. In a market economy individual consumers do not negotiate separately for each unit of a commodity they purchase; rather, they buy as much or as little as they wish at the going price. It follows that the price which the consumer would have been willing to pay for the first pint in our beer example is much greater than the price actually paid; he enjoys a surplus. Similarly, but to a lesser extent, he pays less than he would be willing for the second pint. Adding all the differences between prices paid and those which the consumer would be willing to pay for the successive units purchased (derived from marginal utilities and shown by the demand curve) gives the amount of total *Consumer Surplus*. On Fig. 2.13 it is indicated by the area between the price paid and the demand curve. If the price of beer falls, the amount of consumer surplus rises and if beer prices go up, consumer surplus is squeezed. This concept is often used when analysing the impact on consumers of agricultural policies which raise the price of food to consumers; a reduction in the amount of consumer surplus is interpreted as making consumers worse off.

The Stable and the Unstable Equilibrium

When a consumer has allocated his spending power in the way which gives him the greatest level of satisfaction attainable at the given level of income and prices, he is said to be in *equilibrium*. Furthermore, it is a *stable* equilibrium because, if some temporary factor such as rationing forces him to buy some other combination of goods, when these factors are removed, he will tend to return to his original combination, driven by his desire to maximise his satisfaction. An analogy is a ball in a basin; if displaced to the side, it will return by gravity to the lowest point to restore the status quo. These examples are of stable equilibrium because the forces called into play by disturbing the stable state act to restore the stable state.

An example of an *unstable* equilibrium is a coin balanced on its edge; undisturbed it will stand on its end for ever, but a push will upset its balance and it will fall over because no forces exist which tend to restore the initial position.

Economics is closely concerned with equilibria. Possibly the most important example is the equilibrium which results when the demand for goods and services meets the supply of these goods and services. A balance between demand and supply is achieved through the price mechanism and this is the subject of the next chapter.

Appendix:

A: To show that, where prices of units of goods are *not* the same, maximum satisfaction is obtained when the ratio of marginal utilities (MU) is the same as the ratio of prices.

1. Maximum satisfaction is obtained from a given level of expenditure on two goods when the last penny (or smallest unit of expenditure) spent on either good gives the same amount of satisfaction. This is an example of the Principle of Equimarginal Returns.

2. Where goods cannot be bought in units of one pennyworth, an

approximation of the MUs of the last pennyworth is given by the following:

MU of 1 penny spent on good A $= \dfrac{\text{MU of 1 unit of good A}}{\text{Price in pence of 1 unit of A}}$

Similarly for good B

MU of 1 penny spent on good B $= \dfrac{\text{MU of 1 unit of good B}}{\text{Price in pence of 1 unit of B}}$

Note that the price per unit of A and B need not be the same.

3. If, at maximum satisfaction

MU of 1 penny spent on good A $=$ MU of 1 penny spent on good B

Then, at maximum satisfaction

$\dfrac{\text{MU of 1 unit of Good A}}{\text{Price in pence of 1 unit of A}} = \dfrac{\text{MU of 1 unit of good B}}{\text{Price in pence of 1 unit of B}}$

Rearranging

$\dfrac{\text{MU of 1 unit of good A}}{\text{MU of 1 unit of good B}} = \dfrac{\text{Price in pence of good A}}{\text{Price in pence of good B}}$

This is usually written as

$\dfrac{\text{MU of good A}}{\text{MU of good B}} = \dfrac{\text{Price of A}}{\text{Price of B}}$ or $\dfrac{MU_A}{MU_B} = \dfrac{P_A}{P_B}$

4. For example, if a jacket (good A) costs twice as much as a shirt (good B), a consumer would be spending his money in the most satisfying way if he got twice as much satisfaction out of his last jacket than out of his last shirt.

B: A Note on "Lumpy" Commodities

The indifference curves in Figs. 2.9 to 12 were drawn smooth because it was assumed that between the whole-litre combinations of beer and milk shown in Figs. 2.7 and 8, each schedule corresponding to a given level of satisfaction, any number of further combinations could have been inserted using part-litres. If however only whole litres were available to the consumer, the "curve" should not have been drawn smooth, but as a series of separated points, as in Fig. 2.14. Tommodities which

Fig. 2.14   *"Lumpy Commodities"*

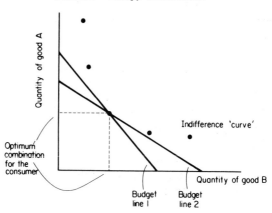

only come in relatively large units are termed "lumpy". Indifference curve theory deals with the situation of "lumpiness" better than utility theory. If a budget line touches the indifference "curve" at a point, it can be seen that quite a large shift in angle of the budget line is required before a new optimum choice results (Fig. 2.14); i.e. relative prices have to alter considerably before the consumer's best choice at one level of satisfaction is changed. If liquids were available to the consumer by the millilitre as opposed to by the litre, the indifference curve would be a much smoother curve because of all the intermediate combinations, and the optimum allocation of spending on beer and milk would be consequently much more sensitive to relative changes in price.

Exercise on Material in Chapter 2

Answer the questions in the spaces provided. Graph paper is given for questions 2.2, 2.7 and 2.8 at the end of the exercise. Please note that in these and all subsequent questions which involve curves drawn by hand some tolerance must be allowed between the answers you obtain and the "correct" answers given in the Appendix.

2.1 In which of the following situations (a) does bread generate utility?
                                        (b) is there a demand for bread?
    Indicate by Yes/No/Don't know.

Utility   *Demand*

(i) A starving man wanting a loaf
of bread.

(ii) A starving penniless man wanting
a loaf costing £10.

(iii) A starving wealthy man wanting
a loaf costing £10.

(iv) A starving man who has an
aversion to bread (the aversion
is total).

(v) A starving wealthy man who has
an aversion to bread (again the aversion
is total. Resale is banned).

2.2 Given the following schedule of total utility for commodity X, derive the corresponding marginal utility schedule. Plot both schedules.

(vi) A man whose hunger has already been fully satisfied.

Units of X consumed per time period	Total Utility	Marginal Utility
1	9	
2	21	
3	35	
4	50	
5	65	
6	79	
7	91	
8	100	
9	105	
10	105	

2.3 The "Law of Diminishing Marginal Utility" states that . . . .
. . . .

2.4 Is the marginal utility schedule derived in 2.2. consistent with the "Law" stated in 2.3?

. . . .

2.5 Give an example of a commodity which could have the characteristics shown by commodity X

. . . .

2.6 A consumer has £20 per week to spend as he wishes on goods A and B. The prices of these commodities, the quantities he now buys and his (subjective) estimates of the utility provided by these quantities are as follows:

	Price	Units currently bought/week	Total Utility	Marginal Utility
A	70p	20	500	30
B	50p	12	1,000	20

Using the above information cross out the inappropriate alternatives in the following statement.

"Assuming that the Law of Diminishing Marginal Utility applies to both goods A and B, for maximum satisfaction this consumer should buy *more/less/the same quantity* of A and *more/less/the same amount* of B. This adjustment of the purchasing pattern will make the ratio of *Marginal Utilities/Total Utilities closer to/more different from* the ratio of *prices/quantities bought.*"

2.7 Below are some of the combinations of amounts of two commodities, X and Y, which correspond to points on three indifference curves for a consumer.

X I_1	Y		X I_2	Y		X I_3	Y
1	50		5	60		10	60
5	30		10	40		15	45
10	20		15	30		20	36
15	15.5		20	24		25	30
20	12		25	20		45	18
25	10		35	15		55	15
55	5		55	10		70	12
70	4		70	8			

(i) Draw these indifference curves (we assume that they are smooth).

(ii) Calculate the Marginal Rate of Substitution on I_1 between X5 and X10.

on I_2 between X15 and X20
on I_3 between X20 and X25

I_1 (X5.X10)	I_2 (X15.X20)	I_3 (X20.X25)
MRS	. . . .	. . . .

(iii) Draw in the consumer's iso-expenditure line (budget line) for each of the income and price levels quoted below, and determine the quantities of the two commodities he purchases in each of these situations.

		Quantities purchased per week	
		X	Y
(a) Income £60 per week	X = £2 each Y = £2 each	. . . .	. . . .
(b) Income £88 per week	X = £2 each Y = £2 each	. . . .	. . . .
(c) Income £110 per week	X = £2 each Y = £2 each	. . . .	. . . .
(d) Income £110 per week	X = £2 each Y = £3 each	. . . .	. . . .
(e) Income £110 per week	X = £2 each Y = £6 each	. . . .	. . . .

2.8 From 2.7 (iii) (c), (d), (e), derive a demand schedule for commodity Y.

Price of Y	*Quantity demanded*
. . . .	. . . .
. . . .	. . . .
. . . .	. . . .

Plot this schedule as a demand curve, with price on the vertical axis.

2.9 Cross out the inappropriate alternatives in the following statement:

"A consumer is said to be in equilibrium when the *Marginal/ Average/Total* Utility from his purchases is *minimised/maximised*. This also implies that the satisfaction he derives from his purchases is *minimised/maximised*. He allocates his spending probably subconsciously to this end by practising the Principle of *Diminishing Marginal Utility/Equimarginal Returns*. The consumer's equilibrium is termed *stable/unstable* since any departures from this position caused by external forces will call into play forces which tend to restore the equilibrium position."

EXERCISE 2.2

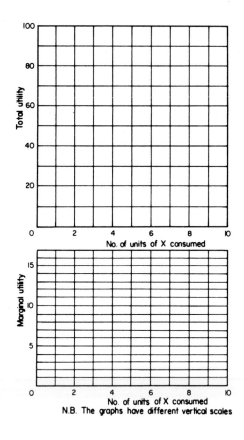

N.B. The graphs have different vertical scales

EXERCISE 2.7

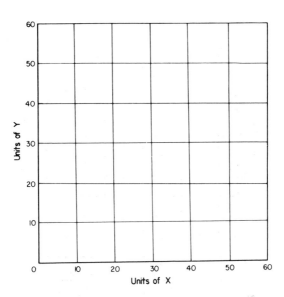

EXERCISE 2.8

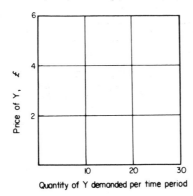

CHAPTER 3

Demand and Supply — the price mechanism in a market economy

Introduction to Demand and Supply

THE study of the demand and supply of goods and services, and the way they interact, forms a fundamental part of economics. Indeed, the majority of economic problems we come across in everyday life can be explained, although perhaps not solved, by a careful examination of the demand and supply of goods or services. It has even been suggested that a parrot could be turned into a passable economist simply by teaching him to say the words "demand and supply" in reply to all questions.

By way of introduction to this important area of study let us take an example from agriculture. At various times of the year some farmers are wanting to buy hay while others are willing to sell hay—a demand for hay and a supply of hay both exist. Hay transactions often occur at auction markets. Let us imagine that we can select one market on one day and have the power to ask as many questions as we like. Furthermore, let us imagine that we can dictate what the price of hay is to be.

If we take all those who wish to buy hay and tell them that hay will be £48/ton that day—a high price for that particular season—very few farmers will wish to buy any and the quantity sold will be small. If we say that the price will be £42/ton, more buyers will be interested. As the price is lowered increasing interest will be shown—more farmers will wish to buy and each man will tend to buy more. If we make a table of the quantity of hay we could sell at various prices we end up with a *demand schedule.*

If we moved on to suppliers of hay we would find that at a low price the quantity which they would be prepared to sell would be small. Only a few suppliers would be prepared to sell at the low price, but if we

Price per ton £	Quantity of hay that buyers are wiling to take (tons)
24	300
30	170
36	90
42	50
48	20

offered a higher figure the quantities offered would increase—more sellers would want to sell and each would want to sell a greater quantity. Again we could draw up a table or schedule of prices and quantities of hay which suppliers would be prepared to supply at these prices.

Price per ton £	Quantity of hay that sellers are willing to sell (tons)
24	50
30	72
36	90
42	104
48	120

We now have two schedules, one of demand and one of supply. If we set the price of hay at £24/ton buyers will want 300 tons but sellers will only be prepared to provide 50 tons so there will be a shortage of hay at that price. If we set the price at £48/ton suppliers will be prepared to sell 120 tons, but only 20 tons will be wanted. Glancing at the two schedules will show that there is only one price at which the quantity demanded and supplied exactly balances—£36/ton when 90 tons will change hands. This is known as the *equilibrium price*.

THE EQUILIBRIUM PRICE OF A COMMODITY IS THAT PRICE AT WHICH THE QUANTITIES DEMANDED AND SUPPLIED IN A GIVEN TIME PERIOD ARE EQUAL TO EACH OTHER.

If we stop dictating prices and give the buyers and sellers free access to one another we will find that, with all the haggling which goes on at these affairs, the actual price paid for hay on that day will settle down at £36/ton. If an auctioneer sold the hay his price would also settle at £36/ton.

In our simple analysis of the hay market, the equilibrium price has been seen as the result of the interaction of the given supply and demand schedules for hay. However, had the weather turned severe early in the winter, the demand for hay might have been greater; this would have forced the price in the market up. Buyers would have been prepared to pay higher prices for the same quantity of hay. If the preceding summer had been very favourable for hay-making, the supply of hay might have been greater, causing a slump in its market price. Had the price of barley been lower, milk producers might have switched to this as a substitute for hay, thereby producing a lower demand for hay and a lower hay price. Hay dealers might have formed a "ring" to fix prices and thus upset the free bargaining between buyers and sellers. Not all hay buyers and sellers are of equal astuteness nor are they equally well informed, and prices of individual lots could be expected to vary around an average. Not all hay is of the same quality and this would also affect the price. Clearly this example as it stands is an oversimplification of what determines prices in the real world, but often only by reducing situations to their simplest elements can explanations be offered for many real-world phenomena. Once a basic model is understood, complications such as those listed above can then be incorporated.

Demand and Supply Curves

It is common practice, and an invaluable aid to comprehension, to express demand and supply schedules in graphical form. When plotting schedules it is conventional to place price on the vertical axis, the curves corresponding to the schedules already given are shown in Fig. 3.1. This is often called the *scissors graph* because of its shape; most demand curves slope downward from left to right—more of the commodity is demanded as price falls—whereas supply curves slope upwards from left to right—more is supplied as price rises. Where the two curves cross is the equilibrium price at which the quantities demanded and supplied exactly balance.

FIG. 3.1   *Demand and Supply Curves for Hay,
Market "X", One Day in December*

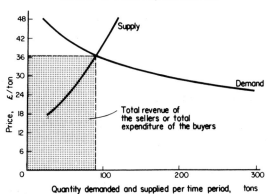

The graph also illustrates the total value of the hay which changes
hands; this value is the same as the total expenditure of all the buyers
of hay taken together or all the revenue of all the sellers of hay taken
together. If the equilibrium price was £36/ton and the quantity supplied
and demanded was 90 tons, £3,240-worth of hay changed hands
(£36 × 90 tons). On the graph this is represented by the shaded rectangle
which is subtended by the point of intersection (i.e. the rectangle bounded
by lines drawn from the point of intersection at right angles to the
axes). It is important to be able to identify this area as we shall be
referring to it later.

Because of the great practical importance of prices to all sections of
the community and to the functioning of the market-based economies
of the Western world, it is worth examining more closely the mechanism
by which prices are arrived at. To do so we will examine separately the
demand for goods, the supply of goods and, thirdly, the ways in which
supply and demand interact.

A. The Theory of Demand

The Meaning of Demand

Demand for a commodity (such as eggs) occurs when people have a desire for the commodity coupled with the willingness and ability to purchase it. A starving man may want some eggs very badly. He may even physically need them to keep body and soul together, but unless he has the purchasing power in his pocket to buy those eggs he will have no impact on the market price of eggs. People who die from lack of food in poor drought-struck countries do so because they lack the resources to buy from parts of the world where it is plentiful. On the other hand, if a farmer has the financial resources which would enable him to buy a new car but he refuses to replace his old one for senti-mental reasons, even though it is in an advanced state of decay, there is no demand from him for a car, in this case through lack of desire. Similarly, religious taboos may eliminate demand for beef or pork irrespective of the spending power of the people who adhere to the religions.

Household and Market Demand

Up to now we have concentrated on individual people and their demands. Often individuals exist in groups—households—the purchases of which are made collectively and reflect the wants and incomes of the various members. A man and wife and (say) two children tend to buy things as a household rather than as four separate individuals. So we talk of *household demand*. When talking about the demand of a whole community for a commodity—such as the demand for milk in the UK—we use the term *market demand*.

If a household has a demand for eggs and buys, say, seven, the demand does not then evaporate, unless its tastes change and it "goes off" eggs. Demand is a *flow*. Eggs are purchased time after time, and if we say a household's demand for eggs at a given price is 7 eggs, we must add *per day*, or whatever the relevant time period is.

Factors Affecting Demand

If we continue with our example using eggs we can show that there are several factors affecting a household's demand for them. Perhaps the most obvious is their price; as the price of eggs rises one would expect fewer to be demanded. But the level of incomes of consumers also bears upon the demand for them; wealthy people may buy more than poorer people; and what about foods that are eaten in place of eggs—breakfast cereals maybe, or eaten with eggs—bacon? What happens to the demand for eggs when the prices of these vary?

The factors affecting household demand for a commodity are
(a) the price of the commodity itself (P_A)
(b) the incomes of consumers (Y)*
(c) the price of competitive (or substitute) goods ⎫
(d) the price of complementary goods   ⎬ (P_B,....,P_N)
                                              ⎭
(e) the tastes of consumers (T)

An abbreviated way of stating that the demand by consumers for good (A) is a function of, or depends on, these factors is

$$D_A = f(P_A, Y, P_B,.....,P_N,T)$$

In addition, when considering what determines total or market demand for a commodity, two other factors must be recognised.
(f) the size of the population
(g) the income pattern of the population

In real life the demand for any commodity is determined by all the variables acting simultaneously. To help analyse how each bears upon demand we must simplify the real-world situation by imagining that all the variables remain constant except for one. By varying that one (say, consumer income) we can focus more clearly on its relationship to demand. This process can be repeated for all other variables singly.

(a) How Demand for a Commodity Varies with the Price of that Commodity

$$D_A = f(P_A, \text{other variables constant})$$

When the prices of most commodities go up the quantities of them which consumers are willing to buy fall. Conversely, when prices are lowered, the quantities demanded increase. An explanation for this was offered in the earlier section on Consumer Choice (see Price Effect on p. 29). Plotting such a situation produces the familiar down-sloping demand curve.

Halving the prices of goods will usually increase the demand for them; for some goods a 50% fall in price will increase their demand

---

* The letter Y is used here to denote income, rather than the more obvious I, because Y is the commonly accepted and long-established label for income when describing the workings of the entire economy (see Chapter 8 on Macroeconomics).

enormously (think what would happen to the sale of Shell petrol if Shell alone cut its prices), but for other goods a similar price cut would make very little difference to the quantity demanded (tap water supplied to farms is a good example). In the first case demand for that brand of petrol is said to be price sensitive or *elastic* and in the second case demand is not sensitive to price, or price *inelastic*. Note that the demand curve for the elastic good (Fig. 3.2) is much flatter than that for the inelastic good (Fig. 3.3).

It is possible to measure the degree of elasticity of demand—its responsiveness to price changes—and the index used for this purpose is the Price Elasticity of Demand.

$$\text{PRICE ELASTICITY OF DEMAND } (E_{Dp}) = \frac{\text{PERCENTAGE CHANGE IN QUANTITY DEMANDED}}{\text{PERCENTAGE CHANGE IN PRICE}}$$

Taking figures from the petrol demand curve: when the price falls from 10 to 8 (a fall or negative change of 2 units) quantity demanded increases from 5 to 9 units (a rise or positive change of 4 units).

$$\text{Percentage change in quantity demanded} = \frac{4}{5} \times 100 = 80\%$$

$$\text{Percentage change in price} = \frac{-2}{10} \times 100 = -20\%$$

$$\text{Price Elasticity of Demand } (E_{Dp}) = \frac{80\%}{-20\%} = -4*$$

To ensure that you know how the calculation was done, work out the Price Elasticity of Demand for water using the figures on the graph. You will find that you get a figure much nearer zero. A coefficient of − 1 means that an x per pent price change causes a change in the quantity demanded also of x per cent. A figure between 0 and − 1 implies that the percentage change in quantity is smaller than the percentage price change (i.e. demand is relatively price-inelastic) whereas a coefficient of more than − 1 (say − 2 or − 3) implies that the percentage quantity change is greater than the percentage price change (i.e. demand is relatively price-elastic).

---

* This figure is accompanied by a negative sign because a movement along a downward-sloping demand curve involves either a negative change in price (and a positive change in quantity) or a negative change in quantity (but a positive change in price). Frequently, however, the sign is omitted and the coefficient of elasticity is taken as the ratio of the proportionate changes.

What Determines the Sensitivity of the Quantity Demanded to Price?

Several factors influence the sensitivity of the quantity demanded to price, reflected in the steepness of the demand curve:
1. The presence of good substitutes.

Shell petrol is a good substitute for Esso petrol in the eyes of most motorists, although individual drivers may prefer one brand or the other. If the price of Shell were increased by a few pence, sales would fall greatly because motorists would switch to alternative petrol brands whose price had not increased and which were now cheaper. The demand for individual brands of petrol is therefore highly price-elastic.

But take a second example, milk. If the price of fresh milk were increased, housewives might buy a little less fresh milk and buy more milk powder. However, for many purposes there is no effective substitute for fresh milk, so if the price were raised by a few pence, the total quantity bought would only decrease by a small amount. On the other hand, if the price were lowered, it is unlikely that much *increase* in sales of milk would occur since there will be little demand transferred to milk from substitutes. The demand for milk is said to be inelastic with respect to price changes because of the lack of good substitutes, and its demand curve is hence quite steep.

FIG. 3.2   *Demand Curve for Petrol*   FIG. 3.3   *Demand Curve for Water*

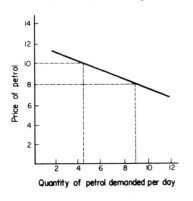
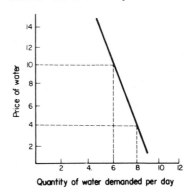

Price elasticity of demand $\left[E_{D_p} \right] = \dfrac{\% \text{ change in quantity demanded}}{\% \text{ change in price}}$

2. The cost of the article in relation to household income.

The typical British household spends very little on salt—perhaps not more than a few pounds in a year. If the price of salt doubled, housewives would probably still buy the same quantity because in their budget it represented an insignificant weekly outlay even at the higher price.

On the other hand, to a student with a car, the cost of petrol represents a considerable proportion of his grant, and he will cut down inessential travel (such as going home to see his parents) if the price of petrol rises.

3. The essential or non-essential nature of a commodity.

It is difficult to get along without some commodities—like basic foods, water, clothing and shelter—so that if the prices of these go up, the quantity demanded will hardly change. On the other hand, it is quite likely that if the prices of essential goods go down, little extra will be taken by the consumer.

4. Habits.

Well-established habits can make consumers' buying patterns insensitive to increases. Indeed, the tax on tobacco takes advantage of this. If putting a tax on cigarettes cut the number sold dramatically, little revenue would be raised. When examined closely many of our purchases are habitual.

Significance of Price Elasticity of Demand (E_{Dp}) to Agriculture

Food in its various forms is agriculture's main product. Foodstuffs taken together have no substitute and are essential to life. On these counts alone we would expect the demand for food to be insensitive to price, giving rise to a steeply down-sloping demand curve and a Price Elasticity of Demand coefficient close to zero. Substitution between different types of food is of course possible and demand for certain items—steak for example—is quite responsive to price changes, but in general quantities demanded are insensitive to price with price elasticities typically between -1 and zero. This is especially true of foods which are not expensive. Some examples are given below.

The steeply downward-sloping demand curve which the agricultural industry faces for its products has important implications for the impact which increased levels of production can have on the revenue to the industry from selling its products, and hence on the incomes of

Commodity	Price Elasticity of Demand (1981-1986)
Fish	− 0.02
Potatoes	− 0.13
Milk and cream	− 0.15
Bread	− 0.26
Fresh green vegetables	− 0.47
Fruit juices	− 0.69
Cheese	− 1.36
Carcase Meat (fresh)	− 1.46

Source: 1986 Annual Report of the National Food Survey Committee

farmers. To illustrate the implications we will use a simple model based on potatoes showing the impact on revenue to potato growers caused by the influence of weather on crop yields and the spread of new higher-yielding varieties; this model contains the essential elements which apply to many agricultural products.

When favourable weather conditions cause potato yields to be heavier than expected, the effect on the supply curve of potatoes is to shift it to the right. At every level of prices growers would be willing to supply a greater quantity than in a year of normal yields (this is anticipating a more detailed study of Supply Theory, but the student should be able to assimilate this idea already). This is illustrated in the accompanying graph (Fig. 3.4) which shows the supply curve for potatoes in a normal year and in a heavy-yield year. The supply curve for the year of favourable weather intersects the demand curve at a much lower price. Note that the point of intersection has moved along the demand curve, which itself has not shifted. At this lower price consumers are only willing to take a little more than they were at the old, normal-year price. Although farmers sell a somewhat greater quantity, this increase is not sufficient to compensate them for the much larger (in percentage terms) price fall. The result is that the revenue which the industry receives (quantity sold times price per ton, represented in Fig. 3.4 by the size of the stippled rectangles) is *less* in the year of heavy yields and with greater tonnage sold than in the year of normal yields. Conversely,

with greater tonnage sold than in the year of normal yields. Conversely, a drought year with lower-than-normal yields would shift the supply curve of potatoes back to the left, greatly raising the price at which the supply and demand curves intersect. Revenue would increase because the fall in quantity sold would be more than compensated by the increased price. We have a situation, then, where the unpredictable influence of weather can cause outputs to vary, resulting in more-than-proportional changes in price, so that higher outputs are associated with lower revenues and vice versa.

FIG. 3.4   *Demand and Supply Curves for Potatoes (hypothetical) Industry level: the effect of favourable weather on the supply of potatoes*

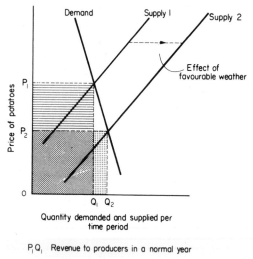

$P_1 Q_1$ Revenue to producers in a normal year

$P_2 Q_2$ Revenue to producers in a year of favourable weather

Consider also what happens when a new variety of potato is developed which yields more heavily, so that for the same costs in terms of land, fertilizer and labour, a greater quantity can be produced. At any given price farmers will be willing to sell more than previously with the effect that the supply curve moves to the right; this is similar to the favourable weather example above but differs in that this rightward movement is permanent whereas that resulting from good weather was temporary and could be reversed. The effect on the revenue which the farming industry receives from potato sales is, however, the same; the total value of potatoes sold is *less* after increasing production through adopting the new variety than before.

Recall that here we are dealing with a model which contains the essential features of a situation, thereby making it easier to comprehend and to make predictions, by omitting many of the real-world complications. For example, we have ignored the effect on total supply which imported potatoes might have in years of high prices and we have not considered the activities of Government-backed organisations such as the Potato Marketing Board in restricting the supply on the market in order to prevent prices falling to very low levels, although the model provides an explanation for such collective action. For some commodities, e.g. milk, these complications may be of major importance in determining prices and revenues. Nevertheless, the insensitivity of demand to changes in price brought about by shifts in the supply curve, which is a characteristic of the demand for many agricultural products, can be seen to be a source of many of the problems faced by the farming industry. These range from the instability in revenues from UK crops like blackcurrants, where relatively minor yield variations can have major repercussions in terms of fruit prices and hence farm incomes, right up to fluctuations in the prices of internationally traded commodities like coffee. Low price-elasticity is also at the root of the long-run decline which can be observed in the prices of agricultural products and plays an important role in the cyles of low and high prices which characterise certain commodities such as pigmeat; both of these phenomena will be re-examined later in detail.

Returning to our potato example, if increasing output reduces the agricultural industry's revenue from potatoes, the logical way for farmers acting together to raise their revenue is to reduce production, forcing the supply curve back to the left, pushing up prices more than proportionally. This is in effect what happens when the Potato Marketing Board restricts the acreage of potatoes which farmers can grow, or when the Board limits the supply to the market for human consumption by diverting some potatoes for stock feed, dying them to prevent their sale to the retail market. The reduction in quantity sold for human consumption is more than offset by the higher prices received. Other UK institutions similarly control output, e.g. the Milk Marketing Board, and supply restriction is an important feature of the Common Agricultural Policy of the European Economic Community. Although farming benefits through higher incomes the consumer has to pay higher prices. This subject is considered at greater length in Chapter 10.

Fig. 3.5   *Demand Curve for the Crop (Potatoes) of an Individual Farmer (hypothetical data)*

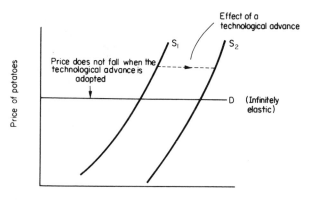

Quantity demanded and supplied per time period

Attempts by the whole industry to restrict output and raise prices will require that individual farmers are prevented from taking advantage of the higher prices by expanding, thereby undermining the collective restriction on output. There will be a tendency for this to happen because, while the demand curve faced by the whole farming industry is down-sloping, *individual* farmers do not face a downward-sloping demand curve for their products. For example, if a farmer uses a new variety of potatoes and has a bumper crop, he will be able to sell them all at the market price and will have a higher revenue than he would selling the fewer potatoes of the old variety. The quantity which he supplies to the market does not affect the price of potatoes because the quantity he has for sale is tiny compared with the total that is being sold. He faces an infinitely elastic (horizontal) demand curve for his products (Fig. 3.5). However, if a sufficient number of farmers simultaneously adopt the new variety, their combined increase in output will push down the market price and hence be felt by the individual growers. Similarly, an agreement between producers to restrict output in order to raise prices and incomes can survive a few "rogue" farmers who run counter to the agreement by expanding, but if the number of "rogues" becomes too great the collective action will be frustrated. A degree of compulsion to comply will be needed, which may take the form of a quota on acreage (potatoes) or quantity sold (milk) imposed by some central authority with fines for exceeding the quota, or contracts with the central body (as for sugar beet) which are, in effect, permits to produce.

Individual farmers will always be looking for new techniques, varieties etc. which increase the quantity they can produce because this increases their revenue. But remember that increases in production act against the interests of the whole industry by greatly pushing down prices. Here is a great dilemma in agriculture—farmers individually trying to expand but farmers acting collectively trying to limit expansion to maintain overall prices.

We must now move to another important factor affecting demand.

(b) How Demand for a Commodity Varies with the Income of Consumers

$$D_A = f \text{ (Y, all other variables constant)}$$

If a household receives a 10% increase in its income it will increase the quantity of goods and services it buys, but it will not increase its expenditure on *all* commodities equally. Very little extra bread will be bought, maybe 21 loaves per month instead of 20 (a 5% increase), but three times as many bottles of wine might be purchased (a 200% increase). Changes in the quantity are related to changes in income by the Income Elasticity of Demand. The formula used to estimate this elasticity for any commodity is:

$$\text{INCOME ELASTICITY OF DEMAND } (E_{D_Y}) = \frac{\text{PERCENTAGE CHANGE IN QUANTITY DEMANDED}}{\text{PERCENTAGE CHANGE IN INCOME}}$$

In the example of bread, when income rose by 10%, purchases of bread rose by 5%.

Hence the

$$\text{Income Elasticity of Demand for Bread} = \frac{5\%}{10\%} = 0.5$$

Similarly, for wine

$$\text{Income Elasticity of Demand for Wine} = \frac{200\%}{10\%} = 20$$

Bread has a low Income Elasticity of Demand and the number of loaves bought is little affected by the change in income. On the other hand, the number of bottles of wine bought is greatly sensitive, and the Income Elasticity of Demand is high. For some commodities an increase in income may result in *less* being bought, and for these Income Elasticity of Demand would be negative. Margarine is such a commodity—as incomes rise consumers switch to other spreads. These goods with negative Income Elasticities of Demand are called "Inferior Goods", a technical term rather than a judgement on their quality.

Figs. 3.6a and 3.6b show that the effect of a rise in income is to shift the demand curves for bread and wine to the right. The demand curve for bread is moved less than that of wine as the demand for bread is the less sensitive to income changes. Note the difference between this effect and that described in the previous section on Price Elasticity of Demand (which does not involve a shift of the curve).

Income Elasticity of Demand; Quantity and Expenditure Measures

The formula given above for calculating income elasticity of demand uses a physical measure of the change in demand caused by a change in income (for example, percentage change in the *number* of bottles of wine bought per week). Occasionally it is useful to calculate the elasticity in terms of the percentage change in *expenditure* on a given item.

$$\text{Income elasticity of expenditure on good "A"} = \frac{\text{percentage change in expenditure on good "A"}}{\text{percentage change in consumer income}}$$

This formula will give same figure as the formula used earlier except where people switch to a more expensive grade of good. For example, a man may still only buy one bottle of wine, but buy a more expensive wine; the quantity will not change but the expenditure will. Strictly speaking, we are not dealing with expenditure on the same

FIG. 3.6a   *The Effect of a Rise in Income
on the Demand Curve for Bread*

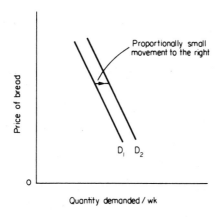

FIG. 3.6b   *The Effect of a Rise in Income
on the Demand Curve for Wine*

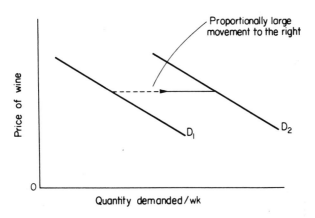

good—cheap wines are not the same as dear wines—but in the real world it may be necessary to lump a collection of dissimilar goods together for the purpose of measurement. Hence published data on income elasticities of demand will usually be labelled as being derived from "quantity" or "expenditure" measures.

Engel's Law

In our example of bread and wine the proportion of the household's expenditure which went on these two items changed when its income increased. Some numbers will make this clear. Say the first income was £10/week, of which £2 (20%) was spent on bread and 30p (3%) on wine. If income went up to £11/week and only 5% extra was spent on bread (10p),* bread would now account for £2.10 out of the £11 of income, or about 19% of income, a fall of 1 percentage point. If three times the original amount was now spent on wine (90p in place of 30p), this would now account for 8% of income—a rise of 5 percentage points. The 19th century statistician Ernst Engel noticed that any additional income tended to be spent more on luxuries and non-essentials than on essentials and his observation, commonly known as "Engel's Law", can be formulated as follows:

THE PROPORTION OF PERSONAL EXPENDITURE DEVOTED TO NECESSITIES DECREASES AS INCOME RISES**

This is represented diagrammatically in Fig. 3.7 which illustrates that expenditure on food and clothes forms a larger proportion of total expenditure of people with low incomes than of those with high incomes.

People's incomes differ and what will be a luxury to a poor person, butter for example, may seem very ordinary to a well-off man, so ordinary that, with the futher increases in income, he switches to low-calorie substitutes which please him better and he buys less butter. A generalised curve relating income to the quantity of a commodity bought (Fig. 3.8) can be constructed and this curve can be divided into four phases.

Phase (a) Income is so low that no butter is bought with increases in income. Income elasticity of demand is zero.

Phase (b) With increases in income more butter is bought; at first a 1% change in income causes a large percentage increase in consumption. Income elasticity of demand is high, but falls as income increases.

Phase (c) Increases in income cause no greater quantities of butter to be consumed. Income elasticity of demand is zero.

---

* 10p is 5% of the original £2 expenditure on bread.
** Engel's Law is commonly restricted to describe the fraction of income spent on food alone and originally took the form of an observation that the poorer the family, the greater the proportion of total expenditure it must use to procure food.

FIG. 3.7   *Income and Expenditure:*
Diagram to Illustrate Engel's Law

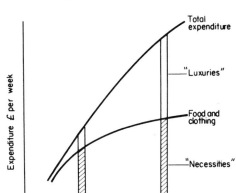

FIG. 3.8   *Generalised Representation of the Relationship*
Between the Quantity Consumed of any Good and
the Level of Consumer's Income

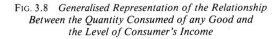
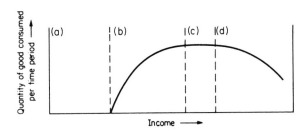

Phase (d) Increases in income reduce the demand for butter—which thus becomes an "inferior good" with a negative income elasticity. Not all goods necessarily become "inferior" at high income levels.

The graph is a general one and can apply to many commodities. For the typical UK inhabitant some goods will be in phase (d) (e.g. margarine $-E_{DY}$ negative), some in phase (c) (e.g. fish $-$ E_{DY} virtually zero), some in phase (b) (e.g. yoghurt E_{DY} positive),* and some in phase (a) (a non-food example—luxury yachts). The income elasticity of demand of all foods taken together is about $+0.23$, which means that food in general lies in the right-hand half of phase (b). Examples of the coefficients (expenditure and quantity purchased) for some UK foods are given below.

Commodity	Income Elasticity of Demand (1986)	
	of expenditure	of quantity purchased
Fresh potatoes	-0.20	-0.32
Margarine	-0.13	-0.26
Liquid milk	-0.10	-0.11
Bread	-0.04	-0.17
Processed cheese	0.14	0.09
Breakfast cereals	0.26	0.28
Carcase meat	0.35	0.27
Fresh green vegetables	0.43	0.17
Fresh fruit	0.70	0.62
Fruit juices	0.94	0.93
All food	0.23	

Source: 1986 Report of the National Food Survey Committee

SIGNIFICANCE OF INCOME ELASTICITY OF DEMAND (E_{DY}) TO AGRICULTURE

Using the E_{DY} of agricultural products we can show why agriculture is a declining industry in terms of the proportion of the nation's income that is spent on it, and consequently in terms of the proportion of the nation's stock of resources (principally manpower) it employs.**

The E_{DY} of food is low compared with that of the products of most other industries. We can construct a simple model to show the relevance of this. Let us imagine an economy made up of only three people—a farmer, a laundryman and another, a consumer called Mr. "N", who buys eggs from the farmer and sends his shirts to the laundry. Mr. "N" has an income of £10/week, of which £1 is spent on eggs and £1 on laundry services. The farmer and laundryman therefore receive the

---

* Figures taken from the 1986 Report of the National Food Survey Committee.

** Between 1900 and 1988 the proportion of the UK's Gross National Product contributed by agriculture fell from 7% to 1.5% and agriculture's share of total labour employment fell from 8% to 2.3%.

same revenue. If we assume that half the revenue of each is kept as income (the other half going on costs of producing eggs or laundering clothes) then the farmer and laundryman will have the same income; they will have income parity. If Mr. "N"'s income goes up by 10%, he will want to buy more eggs and use the laundry more, but it is unlikely that his expenditure will increase on both in the same proportions. Knowing the income elasticities of demand (E_{DY}), we can work out the increases in his expenditure. Assume the E_{DY} of Eggs is 0.20 and the E_{DY} of Laundry Services is 2.00. The formula for estimating Income Elasticity of Demand (E_{DY}) is:

$$\text{Income elasticity of demand } (E_{DY}) = \frac{\text{percentage change in demand}}{\text{percentage change in income}}$$

(quantity *or* expenditure in this example)

Rearranging this:

Percentage change in demand $= E_{DY} \times$ Percentage change in income

Substituting figures:

For eggs

Percentage change in demand (quantity or expenditure) $= 0.20 \times 10 = 2.0\%$

For laundry services

Percentage change in demand (quantity or expenditure) $= 2.00 \times 10 = 20\%$

After Mr. "N"'s rise in income the farmer will find that Mr. "N" will want £1.02-worth of eggs, 2% more than before, and £1.20-worth of laundry services, 20% more than before. A greater expansion in demand has occurred for laundry services than for eggs. The revenue coming to these two producers will have expanded to different extents and, because half their revenue is retained as income, their incomes will have gone up by differing magnitudes. While in *absolute* terms both will be better off, the farmer will be *relatively* worse off because his income will have fallen behind that of the laundryman.

Despite the crudeness of the example, it is a fair representation of what happens in the real world. As a country gets richer, although in *absolute* terms the agricultural industry gets richer, it falls behind other industries and hence *relatively* it becomes poorer. This is not the "fault" of agriculture, or a sign of its inefficiency, but simply the result of the relative income elasticities of demand for its products and those of other industries. The ever-widening gap between the income of agriculture and other industries results in the resources in agriculture—manpower, capital etc.—earning lower returns than elsewhere. There are cries from farmers for subsidies to fill the gap, but this can only be a short-term solution as the gap is bound to widen as a country grows. The long-term solution is to encourage resources in agriculture—and this principally means labour and management—to transfer to other forms of production where higher rewards are available. This transfer will raise the average income of the resources left in agriculture and lower that of resources in other industries, so that parity is again approached. If the shift in resources is rapid, the gap in relative rewards will be small, but larger if they are reluctant to leave agriculture. At the same time, the reallocation of productive factors away from agriculture will reflect better the changing overall pattern of demand for goods and services which accompanies economic growth. This will be re-examined in subsequent chapters.

It is now time to move to the third factor influencing demand for a commodity.

(c) How Demand for a Commodity is Affected by Changes in Price of Competitive and Complementary Goods

$$D_A = f(P_B,....,P_N, \text{ other variables constant})$$

(i) Competitive goods:

Competitive goods are goods which are to some extent substitutes for each other, and they compete to satisfy the consumer's want. An example is butter and margarine. When the price of butter rises, consumers will buy more margarine. Fig. 3.9 shows that a rise in the price of a competitive good shifts the demand curve of margarine (i.e. the curve showing quantities of margarine demanded at different prices of *margarine*) to the right. This happened in the UK when consumers saw butter prices rise as a result of Britain joining the European Community's Common Agricultural Policy. Margarine manufacturers experienced an expansion in the demand for their product; they are also aware that schemes to sell butter at subsidised prices to pensioners could reduce margarine demand.

The sensitivity of demand for margarine to the price of competitive goods, called the Cross Elasticity of Demand, is given by the formula:

FIG. 3.9   *Effect on the Demand Curve of a Price
Rise of a Competitive Good*

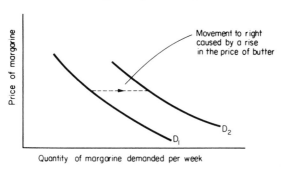

Quantity of margarine demanded per week

CROSS ELASTICITY OF DEMAND FOR GOOD A (E_{D_X})	=	PERCENTAGE CHANGE IN QUANTITY OF GOOD A DEMANDED
		PERCENTAGE CHANGE IN PRICE OF GOOD B

The E_{D_X} of competitive goods is positive, since a rise in price of one will cause *more* of the other to be bought. A rise in the price of the Ford Sierra will result in an increase in demand for Vauxhall Cavaliers; they are strong competitors (or good substitutes) as family and business cars in the UK.

(ii) Complementary goods:

Complementary goods are those which are usually used together. An example is oil and petrol for cars. If more petrol is bought in any one year, more oil is also bought because cars use both together. Such goods are sometimes called "Joint Demand" goods. If the price of petrol rises but that of oil is unaltered, not only will less petrol be bought but less oil too. Cross elasticities of joint demand goods are *negative*. The effect on the demand curve for oil of a rise in the price of petrol is to shift it to the *left*, as in Fig. 3.10.

There is a link between the cross elasticity of demand between two competitive goods and the price elasticities of demand of each good. Goods which are sensitive to prices of competitors will also tend to have shallow demand curves i.e. elastic demand, because consumers will switch rapidly between competitors as one good becomes relatively cheaper or dearer than the other.

Fig. 3.10   *Effect on the Demand Curve of
a Price Rise of a Complementary Good*

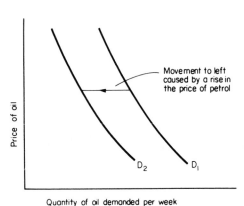

(d) How Demand is Affected by Tastes of Consumers

D = f (T, other variables constant)

Demand has been examined up to now by holding all the influences on demand except one constant, and observing the relationship between the one variable influence and the quantity of goods consumers take from the market. If we held constant the price of the good (P_A), the income of consumers (Y) and the prices of competitive and complementary goods $(P_B,....,P_N)$ there may still be changes in demand, caused by changes in the *tastes* of consumers. An increase in demand caused by a swing in taste towards a commodity will shift the demand curve for that commodity to the right so that more will be bought at any given price. This is shown in Fig. 3.11. The change in taste may be caused by personal experience (e.g. the demand for pizzas in the UK has risen in part through people's experience of Italian food on holidays abroad) by advertising or by a desire to "keep up with the Jones's". This status-seeking consumption shows that the quantity and quality of one consumer's purchases are partly determined by what other consumers do.

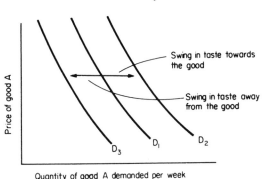

FIG. 3.11   *The Effect of Changes in Taste on the Demand Curve for a Good*

When dealing with relatively short periods of time, economists assume that the consumer tastes remain constant. If they did not, it would not be possible to conclude that, for example, the quantity of butter bought varied with the price of butter, with consumer income and prices of all other goods held constant. Fluctuations in demand *could* be explained by rapid changes of taste and, in some cases are; for example, publicity about *salmonella* contamination of eggs caused a substantial drop in demand during late 1988. Such events are unusual. When dealing with relatively long periods of time, however, changes in consumer tastes *cannot* be ignored.

Additional Factors Influencing Market Demand

At the beginning of this chapter on demand it was pointed out that two additional factors must be considered when analysing the demand by the whole population (market demand) for a good as opposed to the demand shown by individual households. The first is the *size of the population*. A rise in the country's population will increase demand for goods even if average income per head and all prices remain the same. There are simply more people desiring to be clothed and fed etc.; assuming that they possess the purchasing power, the country's demand curve will be shifted to the right. The other factor to be considered is the *distribution of incomes between households.* A change in this distribution will not affect *average* income per head but will cause the demand for some goods to increase and demand for others to fall. For example, if income tax is increased on unmarried men and decreased on married men, leaving average spendable income unchanged, it could be expected that the demand for sports cars in the country as a whole would decline but the demand for family saloons would increase.

Shifts Along the Demand Curve and Shifts of the Whole Demand Curve for a Commodity

To conclude this section on demand, it is worth recapitulating on the ways the factors described above affect the demand curve. A demand curve shows the quantities of a commodity which consumers are willing and able to take from the market at a range of given prices of the commodity. If the price of the commodity is altered, because a shift in the supply curve causes the demand curve and the supply curve to intersect at a different level, more (or less) will be bought; this involves a movement *along* the demand curve. However, the whole curve will be shifted to the left or right by a change in consumer income, a change in the prices of competitive or complementary goods, changes in taste, population changes or a change in the distribution of incomes between households.

Having considered the demand for commodities in detail, the same approach will now be adopted to the supply of goods and services.

B. The Theory of Supply

The Meaning of Supply and Factors Which Determine It

The supply of a commodity can be defined as the quantity that producers are willing and able to offer for sale in a given time period. Like demand, supply is a *flow* of goods and services.

The supply of any good depends on five factors:

(a) the price of the good (P_A)

(b) the prices of other goods which firms could produce or do produce $(P_B,....,P_N)$

(c) the prices of factors of production i.e. the prices of the commodities which are used up by firms $(F_A,....,F_M)$. Factors are supplied by other firms or individuals; examples are electrical power, raw materials, labour

(d) the state of technology (T)

(e) the goals, or objectives, of firms (G).

In abbreviated form, supply can be shown as a function of these factors by the following:

$$S_A = f(P_A, P_B,....,P_N, F_A....,F_M, T,G)$$

The relationships between changes in each of these five variables and changes in supply will now be explored in turn, with the other four held constant. They will be met again in more detail in Chapter 5 when we consider the production decisions of individual farm businesses.

(a) How the Supply of a Commodity Changes with Changes in that Commodity's Price

$$S_A = f(P_A, \text{other variables constant})$$

Normally, as the price of a good increases producers are willing to supply a greater quantity. This is shown by the supply curve in Fig. 3.12; it rises from left to right and hence has a positive slope. (Recall that a demand curve normally has a negative slope and slopes downward from left to right.) Taking eggs as an example, it is reasonable to assume that, if the price of eggs increased, more would be put on the UK market because existing egg producers would expand and some farmers not producing eggs would set up production. Why this happens will be developed in the chapter on production economics; at this stage it is appropriate to accept that producers behave in this way because greater profits can be made by producing more eggs as their price rises.

Fig. 3.13 shows a rather odd supply curve occasionally encountered in agriculture, especially when farmers have invested heavily in specialist buildings and equipment, such as is required in modern large-scale egg production. If prices are lowered to a certain point P, the quantity supplied contracts, but if prices are lowered further, supply *increases.* This can be explained by producers trying to maintain their incomes by producing more as prices fall. For example, a farmer faced with a fall in the price of eggs might cram an increased number of birds into his

Fig. 3.12   *A Typical Supply Curve*

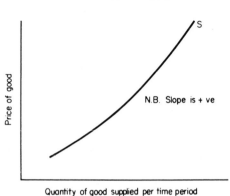

Fig. 3.13   *Supply Curve which is Partially "Reverse"*

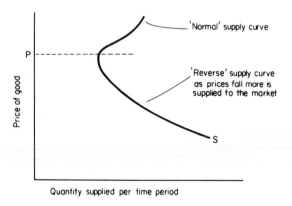

battery house and aim for a higher output. Although profit per egg would fall with a fall in price, by achieving a higher output his overall profit may be maintained or even increased. This type of response, however, must be treated as short-term and exceptional.

Price Elasticity of Supply (E$_{Sp}$)

The responsiveness of producers in terms of output to changes in the price of their product is measured by the Price Elasticity of Supply (E$_{Sp}$). It is defined by the following formula:

$$\text{PRICE ELASTICITY OF SUPPLY} = \frac{\text{PERCENTAGE CHANGE IN QUANTITY SUPPLIED OF COMMODITY A}}{\text{PERCENTAGE CHANGE IN PRICE OF A}}$$

The word "price" is often dropped from the title. Usually the Elasticity of Supply is a positive figure. (It is only negative when the quantity supplied increases when prices fall.)

For some commodities the quantity supplied is extremely responsive to price changes. Such supply is called "Elastic" and the E$_{Sp}$ would be high. Where changes in supply occur without any change in price being necessary, supply is called "infinitely elastic". An example might be an ice cream seller at the seaside; within a given range of quantities he is willing to sell as much or as little as buyers want without changing his prices. At the other extreme, the quantity supplied is very unresponsive, possibly totally unresponsive, to price changes. An example is Cup Final tickets; once the arena has been sold out, however high the "black market" price rises, no more seating spaces can be supplied. In such a situation, supply is completely inelastic.

Fig. 3.14 shows what happens when a change in demand, caused perhaps by an increase in consumers' incomes, meets supply situations of infinitely elastic supply (i), completely inelastic supply (iii) and an intermediate case (ii). With an infinitely *elastic* supply no price increase occurs and the quantity which changes hands increases; with completely *inelastic* supply the price increases but no increase in quantity results, and in the intermediate case *both* the price *and* the quantity sold increases.

What Influences the Elasticity of Supply?

Production is not an instantaneous process and it often takes a considerable time for producers to implement their decisions to change their output in response to price changes. Furthermore, in some cases, while it may be easy to cut back production, it may be difficult to expand so that the responsiveness to price falls may differ from the

FIG. 3.14   *The Effects on Price and Quantity Demanded and Supplied when a Shift in the Demand Curve Occurs*

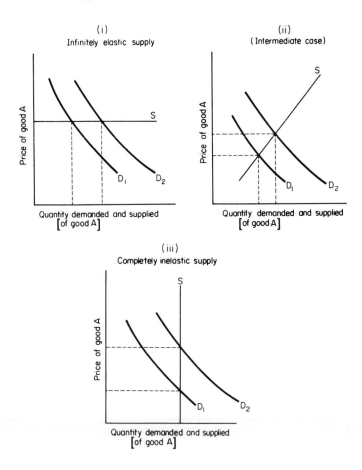

responsiveness to price rises. Hence when considering the factors which influence price/supply responsiveness it is necessary to specify whether expansion or contraction in the quantity supplied is being implied. The influences on the elasticity of supply are:

(i) the time period under review and the length of the production cycle, (chiefly applies to expanding the quantity supplied).

It is obviously impossible to double the supply of barley from UK farms in three months if the price of barley jumps as the result of an increase in demand, so in the short term supply is inelastic. Nevertheless, farmers *can* increase their production of barley dramatically by planting more at the next Spring, so in the longer term the supply is quite responsive to the price and the sharp short-term price rise may be largely eroded. The elasticity of supply will depend on the time scale chosen and the length of the production cycle of the commodity under consideration, which is the time between instigating production and finally selling the product. Taking a second example, it is difficult to increase within a year the production of beef in response to price rises because about two years elapse between the conception of a calf and the slaughter of the finished animal. With cereals the production cycle is shorter and hence production can be expanded more easily—the shorter the production cycle the more elastic is the supply in a given period of time. Taking a third example, the supply of strawberries on any particular day during the picking season is very inelastic. Because they are highly perishable, the fruit has to be sold whatever the price. The overall season's prices will be taken into account in farmers' decisions about production for next year and subsequently; the long-term supply is much more sensitive to price levels.

(ii) the cost structure of production (chiefly applies to reductions in the quantity supplied).

Calves can be reared in expensive specialised buildings with efficient use of feedstuffs, or in cheap rough straw shelters in which much food is wasted; the overall cost per calf may be the same because the differences in the costs of the buildings may be counteracted by the different costs of feedstuffs. However, the composition of the total cost differs and the behaviour of farmers when prices for reared calves fall will depend on which method of calf rearing they are using. The farmer with the cheap straw building may decide to stop production and burn his straw (i.e. with his cost structure dominated by feed costs his supply will be price-sensitive) whereas the farmer with the specialised expensive buildings may decide to "grin and bear" the low prices of his reared animals as long as they cover the cost of feedstuffs. Any margin left over will help pay for the expensive building with which he is encumbered. Because of his cost structure the output of reared calves will not be greatly affected by falling prices, and his supply will be inelastic; he may even try to *expand* production if prices fall.

Another example of the influence cost structure could be expected to have on the responsiveness of supply to falling prices is provided by farms who depend to different extents on hired labour. Many small-scale dairy farms in the UK, on which the work-force consists entirely of the members of the farmer's family, can keep on producing when prices fall (i.e. their supply of milk is price inelastic) because they have no hired labour to pay and are able to put up with lower profits simply by spending less and having a lower standard of living. On larger dairy farms with hired workers forming a large proportion of the labour force, a fall in the price of milk would be followed more quickly by changes to more profitable enterprises. The general decline in the proportion of hired labour in the UK agricultural work-force which has been noted for many years would be expected to make supply of agricultural commodities less elastic, though in practice there may be larger influences moving in the other direction.

(iii) the specific nature of the factors of production (chiefly applies to reductions in the quantity supplied).

The term "factor of production" is used to describe the goods and services which are used up in any production process. For example, the growing and sale of wheat requires land, seed, fertiliser, machinery, men to operate the machines and someone to take the decisions of when to plough, when to plant etc. These factors can be classified in various ways; one way is to group them into factors which can only be used for a single or a few closely related lines of production—termed "specific" factors—and those which can be used in many lines—"non-specific" factors. Seed wheat is a specific factor—it can only produce wheat— whereas most types of land are relatively "non-specific" because a range of crops can be grown, or land can be covered with houses.

Where factors are specific to a particular form of production, that production will tend to continue with a fall in price of the product i.e. supply will be inelastic as prices fall. This is because the factors may have no alternative use, or have only very low earnings in other uses. Earnings in an alternative use are termed *transfer earnings,* because they are what the factor would earn if it transferred. The transfer earnings of a factor are synonymous with the opportunity cost of keeping them in their present employment (see Chapter 2). An example of the effect of low transfer earnings is where a farmer is practically incapable of doing any other job except farming. His transfer earnings would be negligible so he continues farming even when prices drop to very low levels. Another example of a specific factor resulting in inelastic supply is that of the specialist calf-rearing house given earlier.

Elasticity of Supply with Rising and Falling Prices

Stress has been laid on the fact that the supply of a commodity may have a different sensitivity to rising prices than to falling ones. This can produce the odd-looking double-supply curve in Fig. 3.15. This could well describe the situation experienced with products like pigmeat; as prices for pigmeat rise, farmers readily switch to pig production by building up their breeding stock and rapidly constructing new buildings. If prices level off and then fall back as a result of the increased supply reaching the market, farmers will be reluctant to scrap their new buildings and to slaughter their breeding animals. Supply response will be inelastic to price falls until the price drops to a very low level where even feed costs are not covered, when many farmers decide they can no longer endure this level of price. Supply then contracts rapidly with a subsequent rise in the price for pigmeat. Such behaviour patterns by farmers are an important contributing factor to the cycles of low and

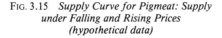

Fig. 3.15　*Supply Curve for Pigmeat: Supply under Falling and Rising Prices (hypothetical data)*

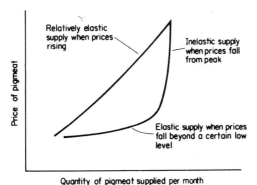

high prices found in pigs, blackcurrants etc.; these cycles will be described more fully later.

It is now necessary to move to the second factor determining the supply of a commodity.

(b) How the Supply of a Commodity Depends on the Prices of all the Other Commodities which are within the Firms' Production Possibilities

$$S_A = f (P_B,....,P_N, \text{other variables constant})$$

If the price of wheat rises, but the price of barley remains the same, the supply of barley can be expected to fall because cereal farms will switch more land into growing wheat. Both crops compete for the farms' resources, and an increase in production of one crop, encouraged by a rise in its price, will necessitate a reduction in production of the other. Hence wheat and barley are termed *competing* products on the farms.

The sensitivity of the supply of barley to the price of wheat is measured by the Cross Elasticity of Supply.

$$\text{CROSS ELASTICITY OF SUPPLY} = \frac{\text{PERCENTAGE CHANGE IN QUANTITY SUPPLIED OF GOOD A}}{\text{PERCENTAGE CHANGE IN PRICE OF GOOD B}}$$

With competing products the cross elasticity of supply will be *negative* .

The effect on the *supply curve* of barley of a rise in the price of wheat is to shift it to the left (Fig. 3.16); at each price of barley farmers will supply less barley than previously because more acres will have been used for wheat and less for barley.

Some products exist whose production process is inseparable from that of some other good; an example is mutton and wool. If farmers are attracted into producing more wool by high prices, more mutton will have to be produced, almost as a by-product. Such closely linked products are termed *Joint Products*; the Cross Elasticity of Supply between such products is positive. The effect of a price rise of wool will be to shift the supply curve of mutton to the right—more mutton will be supplied at the same prices than before the rise in the price of wool (Fig. 3.17).

We have considered in the examples goods which are capable of being

FIG. 3.16   *The Effect on the Supply Curve
of Barley of a Rise in Price of a
Competitive Product (Wheat)*

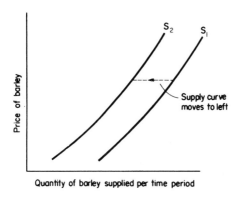

Quantity of barley supplied per time period

FIG. 3.17   *The Effect on the Supply Curve
of Mutton of a Rise in Price of a
Complementary Product (Wool)*

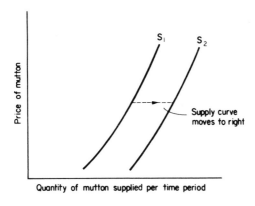

Quantity of mutton supplied per time period

produced by farms, i.e. they lie within the farm's production possibilities.
Prices of goods which are outside the farm's production possibilities will
have no direct influence on the supply of those goods which are inside

its possibilities. For example, the price of watches will have no influence on the supply of pigmeat, because pig farms are not equipped with the necessary machinery or worker skills to produce watches and watch factories are not equipped to rear pigs. The Cross Elasticity of Supply between these two commodities would be zero.

(c) How the Supply of a Commodity Depends on the Prices of Factors of Production

$$S_A = f (F_A,....,F_N, \text{ other variables constant})$$

"Factors of production" are those goods and services which are used in the processes of production. Examples are land, labour, machinery and buildings, raw materials and business acumen. A more rigorous examination of factors of production will be considered later. At this stage it is sufficient to know that each must be paid (or rewarded) for the part it plays, in the form of rent, wages, interest and profit. The first three (rent, wages and interest) will need to be paid by a firm out of its revenue, and anything left over will be the profit for the firm's operators. If the cost of, say, labour rises without the price of the product increasing to compensate, the profit will be squeezed and reduced. The firm's operators may consider that this lower profit is insufficient to compensate them for the risks they are taking in production, and hence may pull out of production. The supply of the commodity reaching the market will fall and the supply curve will shift to the left. The amount of profit that is just adequate to prevent a firm's operator (called the entrepreneur) from ceasing production is called NORMAL PROFIT. If the prices of factors fall, the profits of firms already producing will increase to greater than normal profits. This surplus is termed SUPRA-NORMAL PROFIT. These high profits will encourage new firms to enter this line of production, existing firms will expand, and supply will increase. The effect of a fall in the cost of factors is thus to shift the supply curve of the product to the right, implying that more will be supplied at the original product prices.*

---

* Colleagues perturbed by this somewhat old-fashioned treatment of normal profits are referred to the fuller treatment in Chapter 6.

(d) How the Supply of a Commodity Depends on the State of Technology

$S_A = f\ (T,\ \text{other variables constant})$

In agriculture there is a constant search for new strains of plants or breeds of animals which grow faster or yield more heavily than the existing strains, increasing the output achieved by farms. These changes can be described as technical advances when they enable more to be produced more profitably from a given land area. Total costs to the farmer may rise—new cereal varieties may require heavier applications of fertiliser to realise their potential—but yields rise too, so that the gap between total costs and total revenues, which we can loosely call profit, increases. Another type of technical advance is that which enables the same quantity of a good or service to be produced at a lower cost per unit. An example might be the introduction of a feeding trough in a piggery which reduces the wastage of food so that the same quantity of pigs could be produced at a lower cost. Technical advances which do not also result in an increase in output are comparatively rare. Indeed, many advances *require* an increase in output before their full benefit can be reaped. For example, a new high-powered tractor may only reduce the cost of cultivations if it is used as fully as possible.

The effect of an advance in technology is to shift the supply curve of a good to the right. Since the Second World War the supply curves of almost all crop and livestock products in European countries have moved persistently in this direction, overall by about 2-3% per annum but in some cases much faster. Earlier in this chapter it was shown that the demand for many foods, the products of farming, is relatively inelastic, giving them steeply down-sloping demand curves. An increase in supply caused by adopting the technical advance will depress the product's price so much that the amount of revenue to farmers as a group is actually *reduced*. (Revenue = quantity sold times price.) This was illustrated in Fig. 3.4. In such circumstances, while consumers benefit from lower prices, it is highly doubtful whether the interests of farmers are best served by making use of the technical advance, but the spread of such advances cannot be halted. This is because the first farmers to adopt a new machine or variety of crop benefit most because they can increase their production *while the price of their product is still high.* Prices only start to fall when the bulk of "middle-of-the-road" farmers take up the technical advance. The last, and most conservative, farmers find they are stuck with the old-fashioned way of producing and with falling prices. They, too, are then forced to adapt, by which time prices may have fallen to a level at which the "new" techniques may be no longer profitable. By this time, of course, the most progressive operators have probably moved on to a further advance in technology with yet lower costs and higher outputs. And so the treadmill of technological advance goes on turning. Clearly the secret of making a success of such a changing situation is to be able to recognise early what is a real technical advance and then exploiting it. Not all new machines or crop varieties prove themselves to be advances in technology. Some turn out to be technically interesting but of higher cost than existing farming practices.

(e) How the Supply of a Commodity Depends on the Goals of Firms

$$S_A = f \text{ (G, other variables constant)}$$

It is commonly assumed that the goal, or objective, of any firm is to maximise its profits. However, this is not always the case, and a change in the objectives of a firm can alter the supply curve independent of any change in product prices, factor costs or state of technology. Even when money profits are the only goal there may be a conflict between short-term and long-term profits. A butcher may decide to sell his beef at a lower price than he *could* get in a period of temporary beef scarcity in order to maintain the goodwill of his customers. Without that goodwill he might find his clients drift permanently to other butchers. Firms may be less interested in high profits than in low-risk profits. This is understandable when there is much borrowed capital; a young farmer with a large overdraft may prefer a dairy herd and a regular monthly milk cheque than to risk the overall more profitable but less reliable return from beef production. If the government intervenes to reduce the riskiness of farming, for example by guaranteeing a minimum price for agricultural products, farmers will alter their production patterns. They will specialise more and increase output as a result; they no longer need to take so many precautions to avoid risk.

Where the personal and business lives of the firm's operator are inextricably mixed, as in farming, the running of the farm may be geared as much to having a pleasant life, with sufficient spare time for hunting or racing, as to money profits. This is particularly the case once a certain standard of living for the farmer has been attained. In addition, a farmer may enjoy growing a particular crop or class of livestock; some farmers regard themselves as "sheep-men" and would be unhappy if sheep were banished from their farms in the interest of higher profits. Really they are preferring to take part of their rewards from farming in non-monetary satisfaction. However, if these non-monetary goals are changed in favour of other non-monetary goals or in favour of money profits, the supply curves of the firm's products are bound to be affected and moved to right or left depending on the individual circumstances.

The review of the factors bearing on supply is now complete. Having considered demand and supply separately, it is appropriate to examine how they react when they meet in markets.

Exercise on Material in Chapter 3

Answer the questions in the spaces provided.
Answers and explanations are given in the Appendix.

3.1 The following imaginary demand schedules for onions relate to the three different income groups in a community. Complete the total market demand schedule for the community.

Price per pound p	Rich group	Demand (thousand lb per week) Middle group	Poor group	Total Market Demand
10	4	10	10	
9	5	12	12	
8	6	15	16	
7	6	18	23	
6	6	22	30	
5	5	26	40	
4	5	29	52	

3.2 Cross out the inappropriate alternatives in this statement:

"If the price elasticity of demand for a good is equal to minus 3, then a 1% price fall will *raise/lower* the quantity *demanded/supplied* by *1%/3%*. The total revenue of the sellers of the good will *increase/ decrease/remain unchanged* and the total expenditure of the buyers of the good will *increase/decrease/remain unchanged.* "

3.3 What is the coefficient of Income Elasticity of Demand for housing of a man who, when his income goes up from £2,000 per annum to £2,400 per annum, spends an extra 20% on housing.

3.4 Define Engel's Law ....
....
....
Which of the following are compatible with Engel's Law?
(a) As a country's income goes up, less is spent on essentials.
(b) As a country's income goes up, more is spent on non-essentials.
(c) As a country's income goes up, proportionally more is spent on non-essentials.

3.5 Cross out the inappropriate alternatives in this statement:
"The Elasticity of Supply with respect to product price is estimated as the percentage change in quantity supplied per time period, *multiplied/divided* by the percentage change in *price/quantity demanded*. The Elasticity of Supply is generally *higher/lower* in the short run than in the long run, and this implies that the long-run supply curve is generally *more/less* steep than the short-run curve. In addition, the Elasticity of Supply will be *higher/lower* if the producer has no alternative lines of production open to him; if the production cycle is short the Elasticity of Supply will be *greater/ lesser* than if it is long."

3.6 List (a) three events which would cause a shift in the demand curve for beef, and (b) three events which would cause a shift of the supply curve of beef.

	(a)		(b)
(1)		(1)	
(2)		(2)	
(3)		(3)	

3.7 You are given the following information about the market for carrots:

Price (pence per lb)	Amount demanded per week (million lb)	Amount supplied per week (million lb)
9	30	62
8	35	60
7	41	57
6	45	53
5	49	49
4	53	45
3	57	41

(i) Plot both schedules on the same graph*

(ii) What would be the equilibrium price? . . . .

(iii) What would be the effect of the Government's fixing the market price at (a) 6p and (b) 4p?

(iv) The Government decides to act as the sole buyer of carrots from producers and the sole seller of carrots to consumers. If it decides to fix the price to producers at 6p per pound—

(a) what quantity will it need to purchase? . . . .

(b) what will be its total expenditure? . . . .

(c) what price will it need to charge consumers to clear this quantity? . . . .

(d) what will be its revenue from selling carrots?. . . . .

(e) what will be its net gain or loss from this transaction (ignoring administration costs)? . . . .

(v) Using the original schedules and curves, estimate the new equilibrium price after a change in consumer taste has caused demand to increase by 4 million pounds at all prices.

New equilibrium price . . . .

What quantity would be sold at that price? . . . .

---

* Graph paper with labelled axes is provided at the end of this exercise.

(vi) Using the original schedules and curves, estimate the new equilibrium price after the Government has placed a purchase tax of 1p per pound on carrots. (This has the effect or raising vertically the supply curve by 1p at all levels of supply.)

New equilibrium price ....

What quantity would be sold at that price? ....

(vii) Using the original schedules and curves, estimate the new equilibrium price after the Government has introduced a production grant of 2p per pound. (Producers after receiving grant will be willing to supply at 7p the quantity which they were previously only willing to supply at 9p.)

New equilibrium price ....

What quantity would be sold at that price? ....

(viii) What will be the effect on equilibrium price when an increase in consumers' incomes occurs if:

(a) carrots have a negative income elasticity of demand?

....

(b) carrots have a positive income elasticity of demand?

....

3.8 You are given the following market demand schedule:

Price (units)	Quantity demanded (units)
10	0
8	2
6	4
4	6
2	8
0	10

(i) Plot this schedule on a graph, and measure the slope of this straight-line demand curve. Graph paper with labelled axes is provided.

Slope ....

(ii) What is the price elasticity of demand

   (a) from price 9.2 to price 8.8?   . . . .

   (b) from price 7.2 to price 6.8?   . . . .

   (c) from price 3.2 to price 2.8?   . . . .

Is the following statement true or false?

"The price elasticity of a straight-line demand curve is constant along its whole length."

True/False

(iii) The above estimates of elasticity refer to arcs of the demand curve. The elasticity at any one point can be estimated from the following formula.

$$\text{Price elasticity of demand at price X} = \frac{1}{\text{Slope of demand curve at price X}} \times \frac{\text{Price X}}{\text{Quantity demanded at price X}}$$

Using the formula, calculate the price elasticity of demand at:

   (a) price 2   . . . .

   (b) price 6   . . . .

   (c) Price 8   . . . .

Cross out the inappropriate alternatives in the following statement:

"Where two demand curves of different slopes cross, price and quantity demanded at that price are the same for both curves. At the point of intersection the steeper curve possesses the *higher/lower* price elasticity of demand, and is therefore said to be the *less/more* elastic of the two."

EXERCISE 3.7

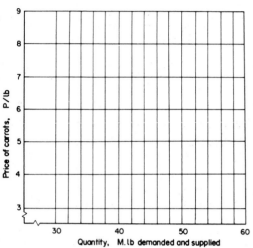

EXERCISE 3.8

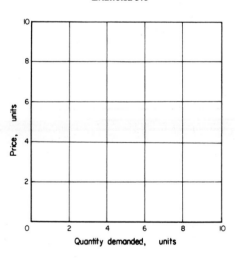

CHAPTER 4

Markets and Competition

Iғ people were entirely self-sufficient, producing the entire range of food, clothes and other requisites from their own resources and the gifts of nature—as might a person marooned on a desert island—then a market in commodities would not exist. Once, however, complete self-sufficiency is left behind and specialisation is indulged in for the greater output that it can generate, with farmers specialising in food production, miners in digging coal, doctors practising medicine and so on, then the problem of disposing of surpluses in exchange for deficits is created. The farmer produces far more food than he and his family require and he will wish to dispose of this in exchange for clothing, heating and other consumer goods. Similarly, a doctor will not be always treating himself and he will wish to sell his services in exchange for non-medical goods and services supplied by others. By this process of specialisation and exchange a higher living standard (implied by the ability to consume more goods and services) is enjoyed than would be possible under self-sufficiency. For most people specialisation takes the form of selling their labour to firms in return for wages. These firms in turn specialise in the production of cars, carpets, crayons, etc. Such specialisation is thus on an organised economy-wide basis, but the aim of increased output and, hence, enhanced living standards is the same. Working for the Government in the armed forces or the Civil Service is just another form of specialisation, although the products of defence or administration are less easily identifiable than is factory output.

The economy thus consists of supplies and demands of myriads of items which must be brought into contact if an exchange economy is to function. The mechanism by which supply and demand interact is called "the market". While it is true that traditionally these interactions between buyers and sellers often took place at specific locations—for example at cattle markets—the term is not now usually so restricted. The cereals market has widened with improvements in communications from local affairs held in "corn exchanges" to as far as the telephone can reach. National and even international cereal markets are concepts easily contemplated, because the actions of buyers and sellers in, say, the USA can affect the price of wheat in Britain.

As we saw in the previous chapter, where a number of willing suppliers co-exist with a number of willing buyers it is possible for a price to evolve at which the quantity that sellers are willing to put on the market is equal to the quantity desired from the market by the buyers. Such a situation would arise when many farmers who are selling store cattle encounter many other farmers who are wanting to buy them at a country town's market. Prices can be agreed either between individuals or through an auction system. Assuming that each store animal is very much like another, if one farmer tries to secure a price higher than the general market level, no one will be interested, and to sell he will need to lower his asking price. Alternatively, if he asks less than the market price he will be bought up immediately, so there is no point in asking a lower price if he can sell at the going market rate. If farmers are unaware of prices elsewhere in the market, they may make transactions above or below the market price, but the better their knowledge of other transactions the smaller will be the price variations. A state of *perfect knowledge* would exist if all buyers and all sellers knew how much each buyer was prepared to buy and each seller was prepared to supply at each price, and only one market price would then exist.

In studying the interactions between demand and supply economists find the use of models of a number of different market arrangements of help in analysing real-world situations and in making predictions. Recall that a model tries to capture the essential elements in a real-world situation while discarding the rest in order to give greater insights into problems than would result from an approach which insisted on taking everything into account. An example is the *Perfect Market*, with "perfect" being a technical description rather than a judgement of what is desirable. This model is of particular relevance to agriculture because many farm products are sold under market situations which contain features closely approaching those of the perfect market model. A perfect market exists when there is:

(a) both a very large number of buyers and a very large number of sellers, the quantity of each buyer or seller being an insignificantly small portion of the total;

(b) a homogeneous product (i.e. the product from any one supplier is indistinguishable from that of any other—a fair example might be barley or milk);

(c) there are no special factors (such as differences in transport costs) to cause any buyer to deal preferentially with any seller;

(d) there are no prejudices in the minds of buyers in favour of or against any seller;

(e) buyers and sellers are free to stop and start buying from the market or supplying the market i.e. unrestricted entry to or withdrawal from the market;

(f) a state of "perfect knowledge" exists.

In such a perfect market only one price would exist at any one time. Any attempts by buyers or sellers to deviate from this price would be thwarted and the price would return to its equilibrium. Both buyers and sellers would therefore be *price takers* in that they could not influence market price by withholding their sales or purchases and would have to accept the going price.

Functions of the Price System

In a perfect market the price system performs a number of important roles. Firstly, in the short run *it matches demand to supply*; for example, while it is possible to buy fresh strawberries in the United Kingdom at any time of the year by importing them, during the British strawberry season the supply to the market is increased vastly. In the short run this seasonal supply is very inelastic (strawberries must be sold or they will deteriorate). To clear the market of this abundance of fruit a fall in price is necessary; at lower prices consumers are willing to buy greater quantities and prices will fall to the level at which demand and supply are again matched.

Secondly, the price mechanism *signals changes in consumer demands to producers*. If consumers take, for example, an increasing liking to beef (the sort of thing which happens when incomes increase) the demand curve for beef will move to the right, resulting in higher beef prices. Beef producers will make high profits; they will attempt to expand production, and new producers will be attracted into beef

Fig. 4.1   *Demand and Supply for Strawberries*

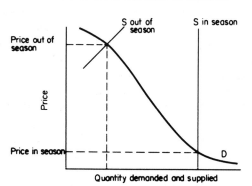

production, eventually increasing the supply. New producers are attracted because existing producers will be making profits higher than those necessary to compensate them for the risks involved in beef production. The lowest level of profit an entrepreneur finds acceptable to compensate him for the uncertainty involved in his type of enterprise is termed *"normal"* profit.* If he persistently earns less than "normal" profits, he will eventually quit production. If he earns more, this "supra-normal" (or "surplus", or "excess") profit will attract other producers; their efforts will increase total supply and lower prices until all "surplus" profit has been eliminated. Producers will then be earning only "normal" profits and no further new ones will set up in production.

Thirdly, and closely linked with the previous point, the price system performs an *allocative function*. Productive resources, controlled by entrepreneurs, are attracted into producing those commodities which the consumer demands. Changes in consumer demands result in a switching of resources to satisfy those changed demands, because they result in commodity price changes to which producers respond. If productive resources are capable of switching from product to product (say, if land is capable of rearing beef animals instead of growing cereals, and farm labourers are as competent with livestock as with crops) resources will be so allocated that consumers will get the goods and services they demand, as indicated by the way they spend their money.

---

* "Normal" profit is discussed in greater detail in Chapter 6.

The fourth function of the price mechanism is to ensure that *goods and services are produced in the most efficient manner.* At this stage we can interpret this as meaning at the lowest average cost. Competition between producers will allow the more efficient to undercut the lowest price at which the less efficient can operate, so the inefficient will be forced out of business. If one size of firm proves to be particularly efficient, then firms larger or smaller will be forced to adopt this size of operation or go out of business, and eventually all firms will be of this optimal size.* The price system thus ensures that the nation's demands are satisfied at the lowest cost.**

Imperfections in the Market

Although the price system plays an important role in allocating the available resources to best satisfy the demands of consumers, it is not a perfect allocator because the best interest of society is not necessarily served by allowing producers to respond solely to the demand expressed by individuals. For example, an unhindered price mechanism would ensure that "hard" addictive drugs would be readily available at minimum cost, being produced in the most efficient manner, whereas society judges (subjectively) that such addiction is neither good for the individual nor for the larger community. In an unhindered system environmental pollution by producers would impose heavy costs on the rest of society, and public goods like the defence forces or the educational system would not be adequately provided. A perfect market would thus not necessarily produce an optimum allocation of the nation's productive resources and Government interference can give a preferable allocation. These "external costs" and "public goods" will be considered in more detail in Chapter 7.

At this stage we are concerned with the imperfections in the price system which occur where individual buyers or suppliers in the market

---

* In farming the continued existence of farms of widely differing sizes is probably a reflection of the lack of marked economies to be earned from greater size.
** "Normal" profit is usually considered as a cost, although not easily quantifiable, and included in total costs. This is because, unless entrepreneurs receive compensation for the uncertainties they face, they will cease production. This compensation, termed "normal" profit, is as essential a payment as labour wages or interest charges; these too must be paid or production will cease. See also Chapter 6 on Factors of Production.

can influence prices for whatever reason. On the supply side, a state of *perfect competition* is said to exist under the conditions laid down for a perfect market earlier (many producers each with an output insignificant when compared with total output, a homogeneous product, unrestricted entry and exit from the market etc.). As soon as individual firms are in a position where they can influence price, imperfect competition is said to exist. A *complete monopoly*, where one firm controls all the output, is the opposite extreme to perfect competition. An example is the GPO monopoly of telephone services. Where supply emanates from a few large firms, such as with farm fertilisers, an *oligopoly* exists. Often supply comes from a mixture of big firms and small firms, and the degree of control over price will then be related to the size of the firm.

On the demand side of the market a *complete monopsony* (note spelling) occurs when there is only one buyer. The Milk Marketing Board for England and Wales is virtually the only buyer of milk from farms in those countries and thus would have considerable power in dictating the price received by farms even in the absence of Government price policy. Similarly an *oligopsony* is where only a few buyers exist—an example would be where there are only a few employers of a specialist skill in a locality.

The imperfections encountered in markets for farm products are not typical of those found in most non-agricultural sectors of the economy. Many non-agricultural goods in developed countries are produced by large firms which are in monopoly positions or are part of an oligopoly, and the danger is that these firms will exploit their power over the market to charge the numerous but individually-defenceless consumers more than they would have to pay under perfect competition. Farming is often faced by the opposite phenomenon; there are many producers of cereals, meat and milk, each of whom are insignificant when compared with total industrial output, but often farmers are dealing with large buyers, such as the marketing boards or large companies, who could at least in theory dictate the prices they offer to farmers. In practice the monopsony is rarely complete, and where it is (e.g. the Milk Marketing Board) farmers usually have an important say in its administration and so the question of exploitation of farmers does not arise.

From the general description of markets given above we turn to examine three models in more detail, firstly, perfect competition viewed from the levels of the individual producer and of the whole industry, secondly monopoly and thirdly monopsony.

The Individual Producer in Perfect Competition

The agricultural industry, consisting of many farmers whose individual outputs form an insignificantly small proportion of total output and whose products are virtually identical to those of other farmers, approximates to the perfect competition model. The individual producer in a perfectly competitive industry, of whom the farmer is a good example, cannot affect the market price for his product. He is a "price accepter" or "price taker". Furthermore, he will receive the same price per unit whatever quantity he places on the market; it matters nothing whether he sells 1 tonne of barley, 10 tonnes or 20 tonnes, the price per tonne will be the same.*

The money received by a firm from selling its products is called its revenue. The revenue from selling a given quantity of output is termed Total Revenue, while dividing Total Revenue by the quantity of output sold gives the Average Revenue, e.g. if the Total Revenue from selling 10 tonnes of wheat is £900, the Average Revenue is £90 per tonne. Average Revenue is therefore the technical term for the price of the product. In the case of the farmer selling barley the Total Revenue he receives will be directly proportional to the quantity he sells and doubling the quantity sold will double the Total Revenue, but the Average Revenue (= price per tonne) will be constant with changing output. This is illustrated in Fig. 4.1a.

Marginal Revenue (MR) is the increase in total revenue to the producer gained by producing and selling one more unit of output, e.g. MR of the 4th tonne of barley is the Total Revenue (TR) from 4 tonnes minus the TR from selling 3 tonnes.

$$MR_n = TR_n - TR_{n-1}$$

The table and graph show that each additional tonne produced increases Total Revenue by the same amount—£50—which is the same as the price (AR) of barley. To the individual producer under perfect competition,

---

* NB. Transport costs are ignored in the model of perfect competition.

then, AR equals MR, and both are constant with increasing output. The AR schedule is also the Demand schedule.

What level of output should the producer aim to produce and sell?* If he is aiming to maximise his profits he will be seeking the largest possible difference between Total Revenue (TR) and his Total Costs (TC) i.e. he will seek to maximise his profit, defined as:

Profit = TR – TC

The output at which this is achieved will generally not be the highest output possible because costs of production must also be taken into account. It is commonly experienced that, once a certain level of output has been achieved, the cost of producing additional units of output rises. An example is barley yields; each additional tonne of barley per hectare requires ever-larger applications of fertilisers, herbicides etc. and hence the cost of each additional unit of yield is greater. The addition to total costs caused by producing the last (marginal) unit of output is called the Marginal Cost (MC) for that level of production.

Marginal Cost (MC)$_n$th unit = Total cost of producing n units
                                                  – Total cost of producing n – 1 units

If the addition to total revenue caused by producing, say, the 4th tonne of barley (i.e. its MR) exceeds the addition to total costs its production entails (i.e. MC), then it will pay the farmer to produce that extra unit of barley because he makes some profit from the unit. It will pay him to expand his output until MC just reaches MR. Beyond this point, further expansion to gain highest possible yields per hectare would *reduce* total profit, because each additional unit of output would cost more to produce than it brought back in extra revenue.

The highest profit will be made when MC = MR. We also know that, under perfect competition, MR = price of the product, so we can state that, under perfect competition, maximum profits will be reaped when output is adjusted so that MC = price of the product (see Fig. 4.2).

In Fig. 4.2 the AR line has also been labelled D (for Demand) and the MC curve S (for Supply); some explanation of this is necessary. The AR curve, which in this example takes the form of a horizontal straight line, shows how the product price changes as the producer supplies

---

* NB. Transport costs are ignored in the model of perfect competition.

FIG. 4.1a   *The Revenue of a Producer under Perfect Competition*

Tonnes of barley supplied to market	Price = Average Revenue (AR) £	Total Revenue (TR) £	Marginal Revenue (MR) £
1	50	50	50
2	50	100	50
3	50	150	50
4	50	200	50
5	50	250	50
6	50	300	50

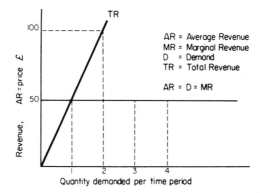

FIG. 4.2   *Marginal Cost and Marginal Revenue of a Producer in Perfect Competition*

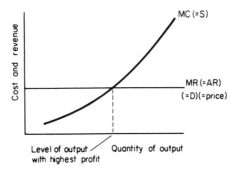

different quantities to the market. But a demand curve also shows how much of the commodity buyers who collectively constitute the market are willing to buy at given prices (see Chapter 3). Clearly an AR curve and a demand curve are showing the same relationship approached from two directions. The demand for barley in this example is infinitely elastic because the market will absorb very little or as much as the farmer wishes to supply, at the same price. Theoretically, if he were to lower his price an infinitely small amount below market price, demand for his product would increase by an infinitely large amount and, if he raised his price above market price, his sales would fall to zero. Turning to the supply label, the MC curve of a firm in perfect competition is the same as its supply curve.* Recall from Chapter 3 that a supply curve shows the quantities which producers are willing to put on the market at different prices. A price fall to our producer in perfect competition would mean a vertical shift downwards of the AR = MR curve, intersecting the MC curve at a lower quantity of output, and indicating that he should produce less. The MC curve thus traces the quantities which the profit-maximising producer will produce and supply to the market at a range of prices, and this is what a supply curve shows.

The Perfectly Competitive Industry

We move now from considering the market situation of an individual producer in a perfectly competitive industry to that of the whole industry. The *supply curve* for a whole industry, made up of many small producing units, can be estimated by adding together the quantities which each firm would be willing to supply at a given price, and repeating this process for a range of prices.

The *demand curve* for the whole industry is not horizontal, as was that faced by the individual firm. As has been pointed out earlier, an industry like agriculture faces a downward-sloping curve for its product because, to encourage greater consumption, prices must fall, and in the case of agriculture, the price fall must be relatively large. The

---

* In Chapter 5 it will be shown that MC can at first fall before the rising characteristic with increasing output, shown in Fig. 4.2, sets in. Strictly the supply curve of a firm in perfect competition coincides only with the rising part of the MC curve for which MC is greater than Average Variable Cost.

downward-sloping demand curve cuts the industry's supply curve at an equilibrium price which the individual producers have to accept. Figure 4.3 should make this clear.

Competition between firms in a perfectly competitive industry will ensure that production is carried out in the most efficient way, as the following explanation shows. As well as his Marginal Costs of production described above, each producer will find that he has a set of Average Costs (Total Costs divided by the quantity produced) which also vary with his level of output, e.g. the average cost of producing barley in £ per tonne. Average and Marginal Costs relate to each other, as shown in the left-hand part of Fig. 4.3; AC is U-shaped, first falling as output is increased and then rising, and the rising MC curve cuts the AC curve at its lowest point.* To continue production, in the long run a producer must receive a price for his product (AR) so that all his costs are covered, i.e. the lowest acceptable AR corresponds with the lowest AC which he can produce at. Those inefficient firms whose lowest attainable Average Costs are not covered by the product's price will soon leave the industry— they are termed "sub-marginal" producers—and only the more efficient will be left in production (see Fig. 4.4). If one size of business happens

FIG. 4.3   *The Individual Producer in a Perfectly Competitive Industry, and the Whole Industry*

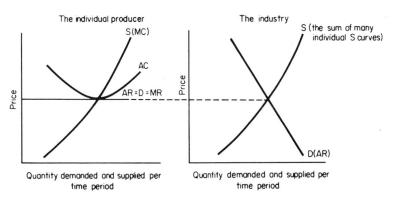

N.B. The price axes are of the same scale for the individual and the industry. The quantity axes will be of different scales - the individual producer's might be in single units, the industry's in millions of units.

---

* This relationship will be expanded in Chapter 5.

FIG. 4.4   *Firms in Perfect Competition*

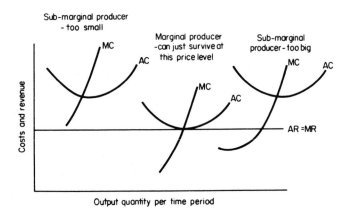

Output quantity per time period

to be more efficient than all other sizes, then firms will move to that optimum size—this will tend to increase the output of the industry, depress market price and force those firms who refuse to adjust out of business. In the long run firms will find themselves producing at the lowest point of their average cost curves—the position indicated in Fig. 4.3—when MR not only is equal to MC (the optimum output position for individual firms) but to AC as well. At this output each firm will be earning just adequate profit to compensate its entrepreneur for the uncertainties of production, but no more. This level of profits is termed "Normal" profits.

A Firm Operating under Imperfect Competition

The next market model attempts to analyse the profit-maximising behaviour of a monopolist in terms of the market situation he faces. Unlike the firm in a perfectly competitive industry, a firm operating under conditions other than perfect competition must face the fact that the quantity it puts on the market will have some bearing on the price it receives. For example, a farmer producing potato seed of a new and spectacularly successful variety, of which he holds the only stock, i.e. he has a complete monopoly, will know that if he only lets a little on to the market he can charge a very high price. If he wishes to sell a greater quantity, his price will have to be lower. His AR, TR and MR schedules might look like this:

Tons of seed offered for sale (i)	Price needed to sell the stock (AR) £/ton (ii)	Total Revenue £ (i) × (ii)	Marginal Revenue £
1	180	180	180
2	160	320	140
3	140	420	100
4	120	480	60
5	100	500	20
6	80	480	− 20
7	60	420	− 60
8	40	320	− 100

Notice that the MR and AR (price) schedules are no longer the same, as they were for the firm in perfect competition back in Fig. 4.1. MR is lower than AR at each level of supply. Total revenue does not go up in direct proportion to quantity sold because the price per ton has to fall. At low quantities, increasing the supply to the market increases total revenue, and MR is positive but falling. Beyond a certain volume the fall in price necessary to sell the potatoes more than offsets the greater volume sold, so that total revenue falls; MR is negative. These relationships are illustrated in Fig. 4.5. It can also be shown that the Price Elasticity of Demand for the product (see potatoes) will be greater than − 1 at quantities corresponding to a positive MR and a rising TR, but will decline with increasing output; it will be − 1 when MR is zero (and TR maximal) and between − 1 and zero when MR is negative (and TR declining).

The farmer in a monopolistic situation will maximise his profits by the same process as the firm in perfect competition—by equating his MC to MR. The important difference is that, because MR and Price (AR) are not the same in this instance, he does not now produce that quantity where MC = Price, but a smaller quantity.* For this quantity the price

---

* Unlike the firm in perfect competition, the monopolist's MC curve cannot be described as being synonymous with his Supply Curve. Depending on the shape of the AR and MR curves, a monopolist might be willing to supply the same quantity at a range of prices and not just one unique price, as under perfect competition.

Fig. 4.5   *Revenue of a Firm in Imperfect Competition*

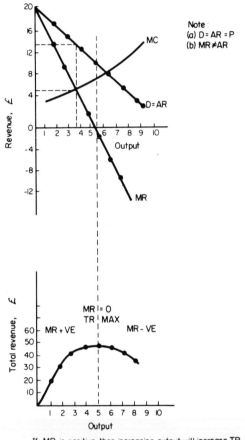

If MR is positive then increasing output will increase TR
If MR is negative then increasing output will decrease TR

charged will be shown by the AR curve; it is clearly above MC while, under perfect competition, price was equal to MC. The implication of the different prices and profits which occur under monopoly and perfect competition will be returned to soon. Note also that at the level of output which maximises the monopolist's profit the Price Elasticity of Demand of his product is greater than -1 (i.e. perhaps -2 or -3).

The monopolist has the choice of either setting a price for his potatoes and letting demand determine how much he sells, or of limiting the quantity and letting competition between buyers determine the price. He cannot fix both price and quantity sold simultaneously.

What will happen if a competitor emerges? If a rival farmer steals seed potatoes and propagates from them, or develops a closely similar strain which is a good substitute, then the pure monopoly is broken. Competition between the established and new firms will cause prices to fall and the profits made by the monopolist will be reduced. To prevent this, and to protect their position, monopolists will discourage the entry of competitors by a wide variety of means—not all of which would be open to our potato producer example. As an established producer, the monopolist will be in a good position to wage a price war against a new entrant by cutting prices temporarily to a level which the competitor cannot withstand because of his initial high costs and low sales. Another method would be for the monopolist to advertise his product heavily so that any potential competitor would need to spend a great deal on making his product known to buyers in order to take sales away from the established producer.

In the real world one frequently hears criticism of the power which monopolies are alleged to wield over defenceless consumers yet, on the other hand, legislation exists to protect some monopolies. Copyrights on published material allow the authors of books and publishers to reap the fruits of their labours by banning cheaper pirate editions until a specified time after the author's death. Similarly, patents protect inventors and trade marks on branded goods can be registered. From this it may be concluded that monopolies are the butt of criticism and yet may not be all bad.

Arguments Against Monopoly

The case against monopolies is based largely on the theory that they result in the consumer being exploited; he is forced to pay a higher price than under perfect competition and purchases a smaller quantity. The "exploitation" argument follows logically if one supposes that an industry in perfect competition, producing as efficiently as possible, is taken over instantaneously by a profit-maximising monopolist. With such a take-over the production costs of each unit (firm) would not

Fɪɢ. 4.6   *Take-over of an Industry in Perfect Competition by a Monopolist*

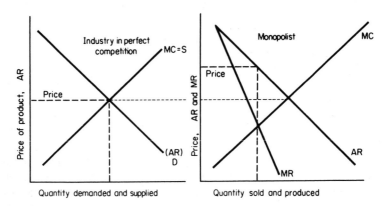

be altered; only the individual entrepreneurs would be rolled into one, or "concentrated". The implications of this are shown in Fig. 4.6. When the monopolist takes over he becomes aware of the difference between AR and MR and to maximise profits reduces output until MC = MR. This difference was not apparent to the former independent producer in the perfectly competitive industry as, whatever quantity he produced, the price he got did not vary. (AR was the same as MR and both were constant with increasing output.) The straight replacement of many independent entrepreneurs by a monopoly is to the consumers' disadvantage, as they now have to pay a higher price and consume less as the monopolist restricts output. The monopolist does this because, although average costs rise as output is reduced, average revenue rises even more, so that profit (revenue minus costs) increases.

Under the perfect competition which existed before the take-over by the monopolist in our simple model, production took place in the most efficient way because in the competitive process high-cost inefficient firms were undercut, with the result that they earned less than normal profits and ceased production. "Normal" profit—that profit required to compensate entrepreneurs for the uncertainties they must face—is usually considered as a cost because, without it, entrepreneurs will stop producing in much the same way as the wages of workers must be met or production will cease. Each remaining firm was producing at its minimum-cost output, so that consumers were paying the lowest prices possible with the techniques of production then known. The price received from customers (AR) just equalled average costs of production (AC including "normal" profit). Yet each firm was earning a profit sufficient to compensate its entrepreneur for the uncertainties of production. If the monopolist who displaces the state of perfect competition restricts output to a level where price (AR) is above average costs of production (AC), it follows that he is earning a *higher* reward than would be necessary to compensate him for the uncertainties he faces, because this compensation element has already been taken into account when calculating AC. This profit above "normal" profit is termed "monopoly profit" and is shown by the labelled area in Fig. 4.7. Obviously the monopolist will attempt to obtain this profit by restricting supply if he is profit-motivated, and consumers will be equally anxious to prevent monopoly profits being extracted from them. However, we shall now show that the case against permitting monopolies is not so cut and dried as our simple model would suggest.

Arguments for Monopoly

An instantaneous process as described above is far from what normally happens in the process of monopoly creation. When a producer develops monopoly power in an industry it is usually a gradual process based on

FIG. 4.7   *Monopoly Profit*

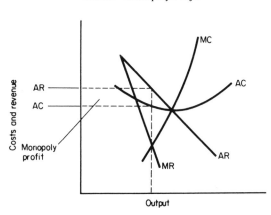

the ability of larger firms to undercut the prices of smaller ones because of the cost advantages of large-scale production. The sources of these economies of scale (technical economies, marketing economies etc.) are discussed in Chapter 5, but as a preliminary the following are among the more obvious advantages associated with a monopolist producing on a larger scale:

(a) monopoly makes possible a more precise *planning* of production. The reduction in uncertainty which results from the elimination of competitors means that monopolists can utilise the best scale of production and the least-cost combination of inputs. Underused units can be shut down and output concentrated in the most efficient way, with a saving in cost. Think of the advantages which might flow from combining two competing but underused animal feed-stuff factories in a county;

(b) a more *economical distribution* of the product becomes possible— in our feedstuffs example only one team of salesmen would be needed, not two competing teams, and one fleet of lorries could be organised in such a way that wasted journeys, due to rival lorries covering the same route, each only partly loaded, could be eliminated. With "public utilities" (gas, electricity, telephone etc.), the duplication of fixed equipment by competitors would most likely be undesirable and costly (several sets of telephone wires or gas pipes under each road etc.). Under a monopoly these "public utilities" can be distributed more cheaply;

(c) a monopolist may be more able and willing to undertake *research* in pursuit of improved production techniques and new products because of the size of research organisation he can support and the sure knowledge that he will be able to enjoy the fruits of his labours undisturbed. On the other hand, critics of monopolies would allege that the removal of competition dulls the spur for research.

If major economies are achievable through arranging production in larger units, then these will offset and maybe exceed the price-raising effect of a monopolist reducing output to exploit his monopoly power (see Fig. 4.8).

There are obviously both positive and negative aspects of monopolies, and instances may occur where monopoly is, as well as being to the monopolist's advantage, also in the best interests of the consumer.

FIG. 4.8   *The Replacement of an Industry in Perfect Competition by an Efficient Monopolist*

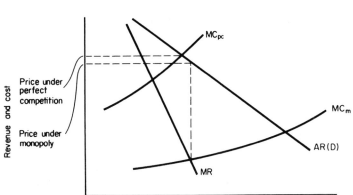

MC$_{pc}$ = Marginal cost curve of an industry in perfect competition

MC$_m$ = Marginal cost curve of an efficient monopolist.

Other instances may exist where monopoly power is clearly exercised against the consumer, making control by society over these monopolies desirable. Some, like the monopoly held by oil producing countries over the world supply of this valuable energy source, cannot be effectively countered except by the development of oil substitutes.* Other industries, such as telephone, gas, water, railways, mining and others without close competitors, were organised in the UK from the 1940s to the 1980s as State monopolies to achieve economies; public ownership of them was intended to ensure that at the same time customers were not exploited. During the 1980s some of these (most notably telephone, gas, some Post Office services) have been transferred to the private sector (with water and electricity to follow) in pursuit of greater efficiency, but with restrictions of their pricing behaviour to protect the consumer.

---

* Some commentators would argue that the monopoly of oil supplies by a few countries and their policy of raising prices is beneficial to the world in general, in that it slows down the depletion of this non-renewable resource hence protecting the interests of consumers in the future.

Where privately-owned industries become dominated by a single or few firms, legislation is available to ensure that they do not exploit their monopoly power excessively (anti-Trust legislation in the USA and Monopolies and Mergers Commission in the UK).

Use of Monopoly Power in Agriculture

We have already noted that one method of arranging support for the incomes of people engaged in agriculture is to force up prices by restricting the supply reaching the market (Chapter 3). While individual farmers cannot hope to affect price by withholding their output, if they act collectively price increases are attainable. Producer-controlled Marketing Boards are such collective organisations; these can take the broad view and analyse the demand situation as any monopolist might. For example, a potato marketing board, if it were entirely under the control of potato growers, might attempt to restrict the supply of potatoes, probably by giving individual growers quotas which they must not exceed, to that level where the industry's Marginal Cost curve cuts the industry's Marginal Revenue curve. To be successful, such efforts would require the banning of imported potatoes (an effective competitor) and strict policing of growers; it would always be in the interests of an individual grower to produce more as prices rose due to the *overall* contraction of supply. Competition from other foods which are acceptable substitutes for potatoes (rice, bread, swedes etc.) would also weaken a marketing board's monopoly position.

When the profit-maximising monopolist was discussed earlier, it was stated that the price-elasticity of demand for the product at the "optimum" level of output (where MR = MC) was above unity (− 1). The actual figures for most foods where marketing boards are involved are below − 1 and for potatoes is only − 0.13.*. The premature conclusion might be drawn that farmers are not acting collectively to exploit their monopoly power over the supply of food as much as they could. In practice, it would appear unlikely that farmers acting collectively could succeed in seriously restricting supply because such attempts, as well as being erodable by dissidents in their own ranks, by imports and by substitutes, would be faced by consumers as unpopular and politically unacceptable price rises.

* 1986 Annual Report of the National Food Survey Committee.

Discriminating Monopoly

A discriminating monopoly exists where a seller can sell an identical article on two (or more) markets at different prices in order to raise its income level. This is of particular relevance to the activities of the UK Milk Marketing Boards which sell milk for liquid consumption at one price and identical milk for cheese and butter manufacture at a much lower price. Manufacturers are prevented by legal restraints from reselling milk to consumers at the higher price. Other examples of this type of behaviour are exhibited by Gas and Electricity Boards (different charges for domestic consumers and industry), compact disc manufacturers who re-issue old recordings at a lower price (retailers cannot buy re-issues to compete with the full-price CD until a time has elapsed and the original deleted from the catalogues), and "cheap" day returns on the railway.

The object of this discriminatory activity is to increase revenue above the level attainable without discrimination. For discrimination to be effective the two markets must have different demand characteristics (otherwise the exercise is pointless) and must be capable of being isolated from each other, by legal or other action, so that "seepage" between markets is not allowed. In the case of milk, the demand for liquid consumption is relatively insensitive to price, giving it a steep demand curve, whereas the demand for milk for manufacturing is much more price-sensitive, with a shallower slope to the curve (see Fig. 4.9).

FIG. 4.9   *A Discriminating Monopolist*

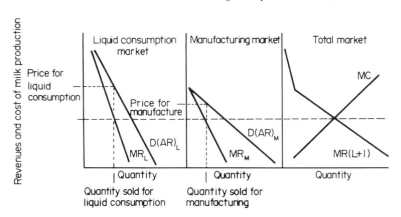

In order to make maximum profit, the discriminating monopolist, the marketing board, should equate MR with MC. However, it realises that its MR curve is made of two components—the MR from sales for liquid consumption and the MR from sales for manufacturing. If it is to maximise its revenue, it must ensure that the MR in each sector of its market is the same, i.e. that the last gallon sold for liquid consumption brings in the same amount of extra revenue as the last gallon sold for manufacturing. This is another example of the Principle of Equimarginal Returns, first encountered in Chapter 2 when referring to consumer equilibrium. In order that MRs in each market can be equated, different prices (AR) must be charged for liquid consumption and for manufacturing, as will be clear from Fig. 4.9.*

A Monopsonist—The Single Buyer

We have considered the effect which a monopolist, either one firm or many firms having a marketing agreement with each other (often termed a "cartel"), can have on a market. Next we must briefly mention the effect of a single large buyer—a monopsonist. Once again using a simple model we pose the question of what happens when, in a perfect market, a large number of *buyers* band together and buy collectively. Such an occasion would arise when a large number of independent small dairies, hitherto competing against each other for the supply of milk from farms, merge into a single buying unit. Let us call this unit the "United Dairy". The supply curve of milk from farms is not altered by such action—the relationship between the quantities farms are willing to supply and the prices at which these quantities will be supplied is not affected by the nature of the buyer.

In order to call forth a larger supply of milk from farmers, "United Dairy" will have to raise the price it offers to farmers and, as a monopsonist it becomes aware of a difference between the average cost of buying milk and its marginal cost, i.e. the increase in total costs it faces as a result of calling for more milk. This was not apparent to the former

---

* In practice a marketing authority may not be concerned with equating MC with MR. It is likely to face the situation of having a given quantity of product and having to dispose of it in the most profitable way. It will still achieve this by equating MRs in its several markets, although the level of this MR may not be of its own choosing.

independent dairies, as each could always buy as much as it wanted at the market price. For example, if the "United Dairy" wishes to increase its supply from farms from 10,000 gallons to 11,000 gallons, and to achieve this it has to increase the price paid to farmers from 45p to 50p per gallon, the total cost of buying milk will rise from £4,500 (45p × 10,000) to £5,500 (50p × 11,000). The marginal cost of the extra 1,000 gallons is £1,000 so that at £1 per gallon it is considerably more expensive than the average 50p per gallon. Armed with this knowledge the "United Dairy" will come to a different buying decision from that which a large number of small, independent price-accepting dairies would reach. This is illustrated in Fig. 4.10a which shows that the monopsonist will buy less milk and pay a lower price than would be the case without a monopsonist. If the supply of milk were very inelastic, which is usually the case in the short run, the quantity taken by the monopsonist would fall only a little, but the farmers would be paid a much lower price (Fig. 4.10b).

Several agricultural products are bought under either complete or largely monopsonistic conditions—milk, sugar beet, blackcurrants—where the market is dominated by a single large buyer. However, most of these monopsonies have arisen through attempts to regulate the markets in those products largely for the benefit of producers and to function as monopolists in *reselling* the product—hence the Milk Marketing Board is a monopsonistic buyer of milk but also uses its power as a monopolistic seller of milk to raise the revenue which producers receive by operating a discriminating monopoly. Where the monopsonist is a producer-oriented marketing board the question of exploitation of farmers does not arise, but it is conceivable, although unlikely to happen, that if the control of such boards were to pass entirely into the hands of consumers, their role might well change as they exercised monopsony power against farmers. If farmers could find no alternative outlet for their goods, the existence of which would undermine the monopsonist's power (such as selling milk direct to consumers) a banding together of farmers might occur to create a monopoly of supply to counter the monopsony of purchase. The result of a tussle between a monopoly and a monopsony (a situation sometimes termed "bilateral monopoly") is not precisely predictable by economic analysis—the outcome is as indeterminate as the haggling between two millionaires over a fine painting. Psychology, politics and countless intangible and unpredictable factors are heavily involved.

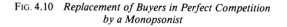

Fig. 4.10 *Replacement of Buyers in Perfect Competition by a Monopsonist*

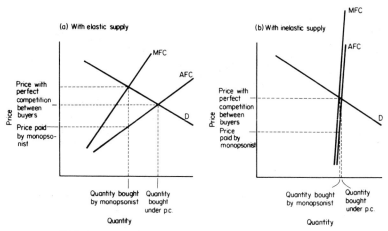

N.B. The D curve in this case is derived from the Marginal Revenue (value) Product of milk to the dairy. The monopsonist is thus behaving in a normal profit-maximising manner by equating MFC with MRP, i.e. buying that quantity of milk at which the cost of the last gallon purchased just equals the extra revenue gained from selling that gallon. This principle will be further developed in Chapter 5

Price Movements in Agriculture

It is time to return to areas of price theory which are of immediate interest to farmers. Often they are aware of periods when prices are depressed and of others when prices are above the anticipated levels. They also have the impression that a long-run downward trend in prices for their products has been occurring. Part of the difficulty in interpreting and anticipating price movements is that the price mechanism reflects a number of changes which are different in cause but which are occurring simultaneously. This is illustrated in Fig. 4.11. Four separate types of price movement are illustrated—the long term trend, medium period, seasonal and daily movements—and the price of a good or service at any one time will be determined by a mixture of all these. Medium period movements can be further divided into those with prime causes lying outside the market involved and those for which the market alone is responsible.

FIG. 4.11 *Price Movements*

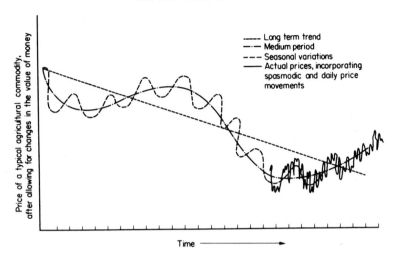

The *long-term trend* in agricultural prices (after having allowed for changes in the value of money) has, historically, been downwards in the developed countries. This is because the expansion in supply of agricultural products has been greater than the expansion in demand. Their supply curve has shifted to the right faster than the demand curve, resulting in a fall in prices (see Fig. 4.12). The population has increased in number (increasing the demand for food approximately in the same proportion) and also enjoys higher incomes, but only a small and declining proportion of this increased prosperity is expressed in terms of extra food demand. (Refer to the section in Chapter 3 on the low income elasticity of demand for food.) On the other hand rapid technological advance (new varieties, new machines, new fertilisers) have meant that increasing outputs have been possible from the existing resources of land, capital and manpower in agriculture. Manpower has even been able to leave farming for other industries and yet farm output has increased.

The implications of this trend are that:

(1) consumers have paid a progressively lower real price for the raw material component of their food (i.e. the basic commodity

FIG. 4.12   *The Long-run Trend of Food Prices
in Developed Countries*

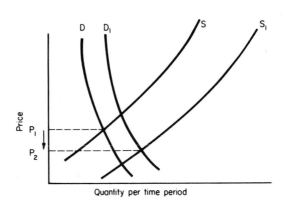

Quantity per time period

excluding processing, packaging and other convenience aspects
reflected in the price);

(2) lower rewards have been earned by resources in farming as
product prices have declined, putting pressure on these resources,
particularly labour, to shift to other more profitable industries
(see the section in Chapter 6 on Mobility of Factors of Production);

(3) strong political pressures have arisen from those who have an
interest in agriculture (notably farmers and landowners) for the
Government to aid the industry by supporting prices against the
long-term trend (see Chapter 10).

The downward trend in prices and the resulting pressure to transfer
some of the resources in agriculture to other industries is the way the
price mechanism works in its role of allocating the nation's productive
resources among their various alternative employments in a capitalist
free-enterprise economy. The market for agricultural products and the
market for agricultural inputs (labour, land and capital in its various
forms) are linked. Rapid increases in farming's productivity coupled
with a fairly restricted demand for its products is bound to require an
outflow of *some* resources from agriculture. The price system is
performing a valid economic function if it indicates the direction of
and assists in such movements of productive resources.

FIG. 4.13   *The Effect of an Interruption to the Foreign
Supply of Food on Prices (hypothetical)*

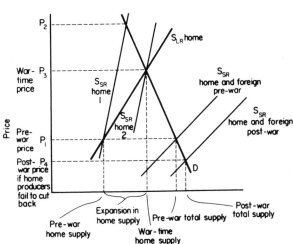

Quantity demanded and supplied

While the consequence of supply outstripping demand poses problems
of agricultural policy for the developed countries, the less developed
countries face the far more serious one of a rapidly-rising food demand
from an expanding population which in many cases exceeds their ability
to produce more food.

*Medium period booms or slumps** are periods of high or low prices,
the causes of which are specific and are not self-regenerating from
within the producing industry. For example, during wartime the foreign
supply of food to a country like the UK is severely restricted. The
supply curve of food to the British market, which comprises both
home-grown and imported food, is shifted to the left as the imported
element is eliminated, raising prices from P_1 to P_2 and causing a boom
for British farmers. In the short run they are unable to increase their
output greatly (their short run supply curve $S_{SR\ Home}$ is steep) but given
time they expand their output as indicated by the less steep long-run

---

* Not illustrated in Fig. 4.11 separately from medium period cycles.

home supply curve $S_{LR\ Home}$ and prices settle at P_3. On cessation of hostilities and resumption of imports British farmers may not immediately contract their production to its original level (in the short run supply is inelastic) so total supply is greater than hitherto and prices drop to P_4, causing an agricultural slump. Slumps can also be caused by an increased availability of imports—this happened when better transport made cheap imported food available to the British market from North America and Australasia at the end of the 19th century. General depressions in the economy can also cause agricultural slumps— consumers coping with reduced incomes may only cut back their demand for food by a small amount but if supply is inelastic the resulting price falls can be considerable.

The long-term price trend is part of the factor-allocation process, and some medium-period price movements can be viewed as performing shorter term reallocation (such as attracting resources into farming to boost food production during wartime). In contrast, *medium period cyclical price movements* which particular agricultural markets themselves foster, albeit unwittingly, do not serve an economic purpose; indeed they can lead to a more wasteful use of productive resources. An example of a cycle is afforded by the regular changes seen in black-currant prices in the UK since the Second World War; prices reach a peak, fall to a trough, rise again to a peak only to fall again. Historically, each peak has been separated by about ten years. Similar but shorter cycles are observable with pigmeat production, beef and a number of other commodities. Cycles are most likely where:

(a) a lag occurs between deciding to produce and the first goods coming on to the market. This means that in the short run supply cannot adjust to price rises, however caused. With black-currants the delay involved might be of the order of 3 years—the time from planting to first significant yield;

(b) entrepreneurs believe that present prices are a good indication of future prices. Present high prices encourage them to launch into preparations for later production by establishing new plantations, oblivious of the fact that similar behaviour by others will so increase the total supply that, when this comes on to the market, prices will be pushed down, causing their decision to produce more to appear much less profitable in retrospect. Once prices have fallen, producers may decide to cut their losses by ceasing production altogether and grubbing their plantations, believing that the low prices will continue indefinitely. Their action, of course, reduces supply and hence raises prices, and the cycle starts again.

(c) the demand for the good is relatively inelastic, so that variations in supply cause marked changes in price. (If demand were infinitely elastic supply changes would not cause any price variations.)

(d) random price variations caused by external influences are commonly experienced—these can initiate and reinitiate cycles. The supply of agricultural goods can be affected randomly by weather. For example, blackcurrants in the UK are particularly susceptible to late frosts; variation in yields can cause wide price fluctuations from year to year. When these spasmodic price changes are necessary to allocate an unusually light crop or to clear the market in a glut year, they serve some economic purpose. They may be counter-productive, however, in that they can initiate price cycles which serve no useful allocative function.

The stages of a typical cycle are illustrated by Fig. 4.14. Price P_{LR} indicates the long-run average price. The dotted supply curve indicates an effect such as adverse weather which raises prices to P_h and initiates the price cycle. Farmers are encouraged to expand their supply up along the supply curve to Q_h, but to induce buyers to take this quantity the market price must fall to P_l. At this low price suppliers are only willing to supply Q_l and resources are taken out of production. However, this reduced quantity results in a higher price, P_h, and farmers are again attracted to switch resources into this line of production.*

Price cycles are considered to be wasteful because of this regular migration of productive resources into and out of a line of production. Farm entrepreneurs never get the chance to allocate their resources in the most profitable, and hence nationally, the most productive way because of constantly changing prices. During high prices resources are

---

* The price cycle will continue if the D-S curves are of the same slope. If D is more elastic than S (i.e. the D curve is the less steep) successive cycles will contract and price fluctuations lessen until the long-run price is reached, although before this happens it is quite likely that the cycle will be restarted by some external influence such as adverse weather. If D is less elastic (steeper) than S, the cycle will continue to expand and price variations widen. The graph of such cycles looks rather like a spider's web and the theory is often termed the Cobweb Theorem.

FIG. 4.14   *A Price Cycle*

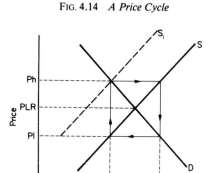

Quantity demanded and supplied

attracted from other lines of production, a movement which would not occur under more stable prices, and so these other lines suffer. Taking an example from the pig cycle, farmers may erect piggeries during high pig prices rather than spend the available money on dairy equipment which could well be the better long-term investment. During low price periods capital assets may be abandoned—disused piggeries put up in times of undue optimism are not an unusual sight in the UK and represent waste of past productive resources.* In addition, consumers do not view gladly fluctuations in the prices of their purchases (bacon, for example), particularly of essentials without adequate substitutes, as it causes fluctuations in their real incomes.

Damping out Cycles

In that cycles cause a less-than-optimal allocation of resources, the whole economy suffers a reduction in its income. It will be in the general interest to prevent cycles becoming established. The enlightened entrepreneur will always make high profits by going against the cycle, timing his production so that it reaches the market when prices are approaching their peak. This may well involve investing in buildings or

---

* All such investments which were triggered off by prices rising above the long-term average might be so considered.

growing stock when product prices are low and other entrepreneurs are withdrawing from production. Wider *education and information* on the existence of the cycle and its causes should encourage more entrepreneurs to take counter-cyclical action, dampening out the price variations.

Another way of obviating price cycles, used particularly in the blackcurrant-growing industry, is for farmers to sell their products *on a long-term price contract* (in this case often for 10 years). This benefits the grower, as he can plan his production with an important element in his decision-making process (product price) settled, and the buyer (principally a drinks manufacturer) is secured of supplies and knows the cost of one of his inputs—the fruit. While economic theory would suggest that the contract price might be somewhat below the long-term average non-contract price (a payment for the reduction in risk involved, see Chapter 6 on "Factors of Production"), this can be offset by the greater efficiency which better organisation based on known prices permits. In times of rapid inflation price contracts may need an adjustment clause to take into account the changing value of money. One other feature of such contracts is that the price cycle may be aggravated for those producers who are not contracted, particularly if contracts are for specified quantities.

A further method of dampening price cycles is for the Government to set *guaranteed prices* for farm produce to provide an element of stability. A simple system of guarantees will not prevent peak prices, but will prevent great slumps, and the average price received will rise (see Fig. 4.15). A *buffer stock* system, whereby the Government buys and stores goods when prices show a tendency to fall below a predetermined level, and puts them back on the market when prices show a tendency to rise rise above a certain level, can prevent this, although such a system demands careful judgement if permanent dumps of goods are not to accumulate.

Seasonal variations in price particularly affect agricultural products, as farming is based on biological growth and not manufacture. Milk in winter will cost more to produce than milk in summer because of the cost of preserving and storing food for the cows. Producers require more revenue per gallon to supply a given quantity—the supply curves for summer milk and winter milk are different, the winter curve lying above the summer one (see Fig. 4.16). Another cause of seasonal

FIG. 4.15 *The Effect of a Guaranteed Price*
on Average Price over Time

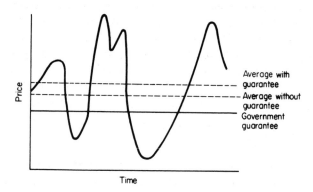

variation in price is seasonal taste changes on the part of consumers: there seems little reason why the supply curve of turkeys at Christmas should be any different from that in other seasons because turkey-rearing is not dependent on the weather and the higher prices at Christmas can be explained by the extra seasonal demand shifting the demand curve to the right (see Fig. 4.16b).

The last price variation to be considered is that relating to the very short term. *Daily* price variations in the free market for strawberries—say, at the Birmingham produce market during July—will be necessary to clear the market daily. An increased supply, caused perhaps by a

FIG. 4.16 *Seasonal Price Variations*

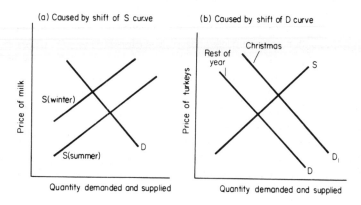

spell of fine weather which speeds ripening and makes picking easy, will tend to push prices down. Unless prices are allowed to slip downwards, goods will remain unsold, not a pleasant prospect with fruit which deteriorates rapidly. With goods which are non-perishable, daily variations in the market price to clear the market will be small or non-existent, as storage is practicable. This can be seen applying in the store animal trade where farmers buy back their own animals if they judge the price to be insufficient and try elsewhere on another day.

The Price System in Centrally-planned Socialist Economies

The price system in a fundamentally capitalist, market-based economy such as the UK or USA can be seen to be the major mechanism by which the nation's resources are channelled into producing the goods and services desired by consumers. The mechanism is not without faults—those emanating from monopoly and monopsony power have been discussed, and others will be described later, including the failure of the market to adequately reflect costs to society such as pollution, stress and the whole host of other environmental factors which accompany a high output/high consumption economy. The Government makes modifications to the market system in an attempt to counteract these faults. Nevertheless, resource allocation derives basically from the market.

In the centrally-planned socialist economies (such as the USSR) the allocation of productive resources is essentially a planned allocation according to what the planners judge should be produced. Such a system can result in an optimal allocation of resources if (a) the preferences of consumers are known; (b) full knowledge exists of the productive resources available; and (c) the technical relationships of production are known. In the absence of this knowledge, misallocation of resources is very likely, and consumers will get what the planners feel they ought to have, rather than what they want. Shortages and surpluses are also likely, because prices are not the result of an interaction between supply and demand (however imperfect) but are planned too, although in recent years an increasing use has been made of markets to indicate the direction of consumer preference. The basis of price-setting is not uniform between products, so prices cannot be used for factor allocation.

Nevertheless, supporters of the planned economic system would put forward the view that planned prices can be adjusted to take into account the social implications of production and consumption. For example, necessities for life, such as food or housing, can be priced low and luxuries high. Demand for consumer non-necessities can be choked off by high prices, leaving productive resources available to build up the country's stock of capital, enabling it to grow rapidly, or to build up a defence force. Private ownership of pollution-generating motor cars can be restricted by making them very expensive.

Such a pricing system supposes that the planners know what is best for the country's inhabitants. In the past the existence of massive queues for some goods in Russian shops and excesses of others suggest that the planners' preferences have not always coincided with the demand consumers have expressed by their buying intentions.

Exercise on Material in Chapter 4

4.1 Cross out the inappropriate alternatives:

In a perfectly competitive industry:
(a) There are *many/few* producers.
(b) Entry to the industry is *free/restricted*.
(c) Each producer *can/cannot* influence market price.

In a complete monopoly:
(a) There is *one/are several* producer(s).
(b) Entry to the industry is *free/restricted*.
(c) The producer has *considerable/no* influence on price.

To a *producer* in a perfectly competitive industry:
(a) Marginal Revenue *is/is not* constant with varying levels of output.
(b) Marginal Revenue *is/is not* identical with Average Revenue.
(c) The demand curve which he faces is infinitely *elastic/inelastic*.

To a *producer* in an imperfectly competitive industry:
(a) Marginal Revenue *is/is not* constant with varying levels of output.
(b) Marginal Revenue is normally *greater/less* than Average Revenue.

(c) The demand curve for his products is normally *down-sloping/ up-sloping*.

A perfectly competitive *industry* normally faces a demand curve which *is infinitely/less than infinitely* elastic.

The Marginal Revenue of a perfectly competitive *industry* normally is *constant/declines* with increasing output.

4.2 A village shop, operating under imperfect competition, finds that by varying the price it charges for rubber boots it can sell different quantities. It finds that it faces the following demand schedule. Derive from this the shop's Total Revenue and Marginal Revenue schedules. (NB. From a seller's standpoint, price is synonymous with Average Revenue.)

Price per pair (Average Revenue) £	Quantity demanded per week	Total Revenue £	Marginal Revenue £
2.75	1		
2.45	2		
2.20	3		
1.90	4		
1.65	5		
1.45	6		
1.20	7		
1.00	8		
0.80	9		
0.60	10		

(a) Plot Average and Marginal Revenue schedules.

(b) The shopkeeper obtains his supply of boots from the wholesaler at £1 per pair. This price does not vary with quantity. Hence the Marginal Cost of boots to the shopkeeper will be £1. If maximum profit is made by the shopkeeper when he equates Marginal Cost with Marginal Revenue, what price will he need to charge for boots to make maximum profits? (i.e. what is the price of boots when Marginal Revenue is £1?).

. . . .

What quantity of boots will be sold at this price?

. . . .

Using the formula given below, estimate whether the price elasticity of demand at this most profitable price is greater or less than unity.

$$E_{DP} = \frac{1}{slope} \times \frac{P}{Q}$$

. . . .

4.3 A country has 50% of its resources employed in agricultural production, and 50% of its resources in other forms of production. Half of the national income of £100M is used in buying goods from agriculture and half in buying goods from the other forms of production.

An improvement in productivity increases the national income to £110M. If the income elasticity of demand for agricultural products is 0.4 and for the products of the other industries is 1.6,

(a) What is the extra that will be spent on agricultural products?

. . . .

(b) What is the extra that will be spent on other products?

. . . .

(c) What will be the final income to the agricultural sector?

. . . .

(d) What will be the final income to the other sector?

. . . .

Use your present knowledge to cross out the inappropriate words in the following statement.

"After the rise in national income the half of the nation's resources in agriculture is earning *more/less* than the half in the other industries. Resources should be transferred *to/from* agriculture *to/from* the other industries if the return to the nation's resources is to be equated in each sector."

4.4 Cross out the inappropriate alternatives in this statement:

"If a firm in perfect competition is attempting to maximise its profits it should produce that level of output where the addition to total costs caused by producing the last unit of output (called *Average/Marginal* Cost of production) just equals *Average/Marginal* Revenue. In perfect competition this will also *equal/not equal* the price of the product. A monopolist *will/will not* achieve maximum profit by pursuing the same policy, and in his case the Marginal Revenue at his optimum level of output will be *greater than/less than/equal to* the price which he charges for his product."

4.5 Cross out the inappropriate alternatives in this statement:

"Compared with a non-discriminating monopolist, a monopolist engaged in price discrimination will generally have a *smaller/ greater/the same* revenue. This is because the discriminating monopolist equates the *Average/Total/Marginal* revenue for his product in each market between which he discriminates."

4.6 A producer of milk on an island finds that he can operate a discriminating monopoly. He can sell milk either for liquid consumption or to the island's cheese factory. By charging different prices for milk for liquid consumption and for manufacturing, the producer finds that he can increase his revenue above the level he would receive if he charged one price.

(a) What are the two chief conditions necessary before a discriminating monopoly can operate?

. . . .

. . . .

(b) His Average Revenue schedules for the two markets are given on the next page. Construct Total Revenue and Marginal Revenue schedules.

The AR and MR schedules have been plotted in the graph accompanying this question, and the *combined* MR constructed by adding the two components. If the producer's Marginal Cost curve cuts the combined Marginal Revenue curve at an ouput of 12 gal/day, to maximise profit

Liquid consumption				Cheese-making			
Quantity sold (gal/day)	Average Revenue (£/gal)	Total Revenue £	Marginal Revenue £	Quantity sold (gal/day)	Average Revenue (£/gal)	Total Revenue (£)	Marginal Revenue (£)
1	0.45			1	0.29		
2	0.40			2	0.28		
3	0.35			3	0.27		
4	0.30			4	0.26		
5	0.25			5	0.25		
6	0.20			6	0.24		
7	0.15			7	0.23		
8	0.10			8	0.22		
9	0.5			9	0.21		
				10	0.20		
				11	0.19		
				12	0.18		
				13	0.17		
				14	0.16		
				15	0.15		
				16	0.14		
				17	0.13		
				18	0.12		
				19	0.11		
				20	0.10		

(c) What price should the producer charge for liquid milk?

. . . .

(d) How much milk will he sell for liquid consumption at this price?

. . . .

(e) What price should the producer charge the cheese factory for milk?

. . . .

(f) How much milk will he sell for cheese-making at this price?

. . . .

EXERCISE 4.6

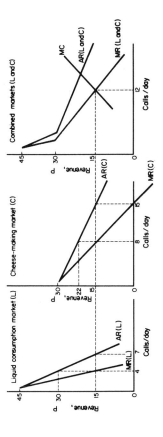

CHAPTER 5

Production Economics
(Theory of the Firm)

NOTE on Chapter 5: this chapter contains essential material for students whose prime interest is business management. However, those with major interests in other areas may prefer to pass on first reading to the last section on "Time and Scale of Production". The omission of the first part of this chapter will not undermine their understanding of subsequent chapters.

The Scope of Production Economics

Production Economics studies how one sector of the economic system, *firms*, allocate their resources in the pursuit of given objectives. Parallels exist between this area of economic theory and the Theory of Consumer Choice already encountered in Chapter 2. When the Theory of Consumer Choice attempted to explain how individual consumers allocated their purchasing power, it was assumed that the goal of any consumer was to maximise his satisfaction. Similarly, in Production Economics the assumption is made that the goal of any firm is the maximisation of profit. However, the two areas of theory differ in that profit, unlike personal satisfaction, is fairly easily measured. (Note that in this chapter the term "profit" is used rather loosely and in an accountancy sense to refer to the difference between revenues and whatever costs are being considered at the time.)

Production Economics recognises that in practice firms do have other objectives (or goals) besides profit. These were discussed in relation to the Theory of Supply (Chapter 3) and included the avoidance of risk, the prestige of the business and the personal preferences of the entrepreneur. For the sake of simplicity, however, Production Economics assumes that the sole objective of production is the maximising of profits. Non-profit motives can be incorporated later.*

The resources at a firm's disposal consist of its funds, which can be spent on machinery, buildings, raw materials, the hire of labour and land or factory space, and its entrepreneurship, or management. The entrepreneur must allocate and organise the other resources so that they are used in the best possible way—that is, so that the highest profit possible is generated from them. Production Economics is often called the Theory of the Firm because it attempts to explain the behaviour of profit-maximising firms and can be adequately described by modifying the general definition of economics given in Chapter 1 of this text.

"PRODUCTION ECONOMICS, OR THE THEORY OF THE FIRM, IS THE STUDY OF HOW FIRMS ALLOCATE THEIR SCARCE RESOURCES BETWEEN ALTERNATIVE USES IN THE PURSUIT OF PROFIT MAXIMISATION."

Questions Facing the Entrepreneur

In the pursuit of profit an entrepreneur will have to make three major decisions: (a) what to produce; (b) how to produce it; and (c) how much to produce. To a large extent the resources he has at his disposal, including his own management preferences and abilities, will answer these questions for him. For example, a farmer with land, labour and a certain stock of machinery and access to a certain amount of borrowing from banks etc. will naturally not consider producing motor cars or television sets because he does not have the right types or quantities of resources required. He will only consider products lying within the range of his abilities and resources.

Given time, the farmer *could* sell his land, stock, machinery etc. and set up as a manufacturer of, say, washing machines if he had the ability to manage such an enterprise. This illustrates the importance of time in any decisions made by firms. Generally, Production Economics is concerned with short or medium time periods in which the entrepreneur

---

* The relative ease with which the objective of production can be measured and hence the success or failure of a management decision assessed, has led to areas of the subject becoming highly mathematical. However, in this introduction to Production Economics, the use of mathematics is minimal.

is unable to change radically the nature of the resources he possesses and in which he is restricted to making small (or marginal) changes to their quantity and their allocation. Examples of problems studied by Production Economics are (a) the best balance between wheat and milk production on a farm, or (b) the cheapest combination of foodstuffs to enable a cow to give a certain yield, or (c) the most profitable yield of cereals (it will not necessarily be the highest possible yield) at given levels of grain prices and fertiliser costs.

Major Relationships Studied in Production Economics

The choice of the best combination of enterprises (or products) on a farm with a given level of resources is termed "optimising the product-product relationship". Choosing the best combination of inputs (factors of production) to produce a given level of output—e.g. the cheapest ration to produce a given milk yield—is termed "optimising the factor-factor relationship". Choosing the most profitable level of fertiliser (input, or factor of production) on cereals (output, or product) given the level of the other inputs, is termed "optimising the factor-product relationship". An entrepreneur in practice is faced with the problem of optimising simultaneously these three relationships. He must select the best combination of wheat and grass (and other enterprises) for his farm and at the same time choose the most profitable yields of milk and wheat etc. to aim for whilst choosing the lowest cost methods to achieve these yields.

Optimising the Factor-Product Relationship

A classic example to illustrate this relationship is fertiliser and crop yield. Assume that a farmer has just planted a 5-hectare field with barley; how much fertiliser should he use to give the most profitable yield? Fertiliser application is the only factor which he can alter to affect yield since all the other factors are fixed in quantity—the size of the field, the quantity of seed he has already put in the ground, his regular labour force (so that he cannot send extra men in to hold up ears that become lodged) etc.

FIG. 5.1   *The Yields of Barley Achieved with a Range of Fertiliser Usages on a Fixed Quantity of Other Factors of Production*

Units of fertiliser per hectare	Yield of barley* (units per hectare) Total Product	Marginal Product	Average Product (nearest whole number)
0	0	—	0
1	6	6	6
2	18	12	9
3	40	22	13
4	76	36	19
5	106	30	21
6	130	24	22
7	150	20	21
8	160	10	20
9	166	6	18
10	170	4	17
11	172	2	16
12	170	− 2	14
13	166	− 4	13
14	160	− 6	11
15	152	− 8	10

* The yields of barley in this example and the data of all other examples in this chapter are hypothetical.

He will find that by using more fertiliser he will get higher yields.* At first the response of the crop is great but beyond a certain level of application the response to additional units of fertiliser is smaller. If very high levels of fertiliser are used, its further use may actually *reduce* total yield because the crop may be poisoned, or grow so succulent that it collapses in the slightest wind.

A schedule of possible levels of fertiliser use (input) and yields of barley (product) is given in Fig. 5.1. The increase in the product caused by the last (marginal) unit of fertiliser at each level of fertiliser use is called the *Marginal Product*, and this is also shown.

---

* The relationship which output (barley yield) bears to the quantity of input (fertiliser) can be described mathematically, and this mathematical relationship is called a *Production Function*. In our example the Production Function in its simplest symbolic form might be written as follows, with "f" standing for "is a function of" or "depends on":

barley yield = f (fertiliser application, all other inputs held constant).

MARGINAL PRODUCT OF INPUT X IS THE INCREASE IN TOTAL PRODUCT PRODUCED BY THE LAST UNIT OF THAT INPUT, ALL OTHER INPUTS REMAINING CONSTANT.

Beyond a certain level of fertiliser use the Marginal Product declines. This is found so generally that it has led to the formulation of the Law of Diminishing Marginal Product (sometimes called the Law of Diminishing Returns).

BEYOND A CERTAIN LEVEL OF OUTPUT, WHEN SUCCESSIVE UNITS OF A VARIABLE INPUT ARE ADDED TO A FIXED QUANTITY OF OTHER INPUTS, THE ADDITION TO TOTAL PRODUCT CAUSED BY EACH SUCCESSIVE UNIT OF VARIABLE INPUT DECLINES.

Other examples of Diminishing Returns can be seen in the overfeeding of animals or in stretching the use of machines beyond their designed capacity such as might occur when an increased number of cows is put through a small one-man milking parlour.

This law is very similar to the Law of Diminishing Marginal Utility experienced in the Theory of Consumer Choice (Chapter 2). As with that Law, diminishing returns (falling Marginal Products) do not necessarily start at once. Fig. 5.1 shows that increasing returns accompany the first few units of fertiliser, but this is soon reversed. It is obvious that, in the real world, Diminishing Returns do occur. If they did not, it would be possible to grow the whole world's food requirements on one hectare of land.

Also in Fig. 5.1 is the Average Product of the fertiliser—this is the Total Product (yield) at each level of fertiliser use divided by the total quantity of fertiliser needed to produce it. Notice that Average Product rises, then falls. When 6 units of fertiliser are used, average yield at 22 units of barley per unit of fertiliser is highest. From a *technical* viewpoint, this is the most efficient level of production to aim for, but it may well not be the level at which most profit is made. *Economic* efficiency is concerned with this most profitable level, so economic and technical efficiency mean different things.

Total Product, Marginal Product and Average Product at different levels of variable input (fertiliser) are shown in graphical form in Fig. 5.2. Notice that:

(a) The point where Marginal Product (MP) starts declining (called the point of maximum Marginal Product, or Point of

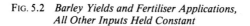

Fig. 5.2   *Barley Yields and Fertiliser Applications,*
All Other Inputs Held Constant

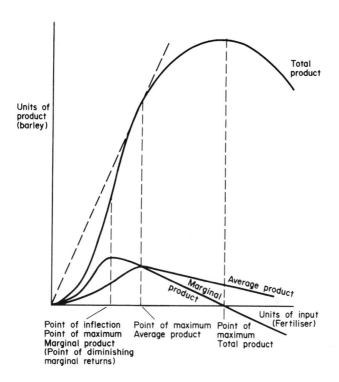

Diminishing Returns) corresponds to the point of inflection of
the S-shaped Total Product (TP) curve, i.e. where with increasing
units of fertiliser input the TP curve stops bending left and
starts bending right. This is easily understood once it is pointed
out that the Marginal Product is the same as the *slope* of the TP
curve, and at the point of inflection the slope of the TP curve
is at its greatest;

(b) MP falls to zero when TP is at its highest because the slope of
the TP curve at this point is zero. Further units of fertiliser
reduce total barley yields; the TP curve takes on a negative
slope and MP becomes negative;

(c) The Average Product at each level of input use is the same as the slope of a line drawn from the origin to the TP curve. The greatest slope this line can have is where it is tangential to the TP curve, and hence AP will be at its highest at this level;

(d) The falling MP curve cuts the AP curve when Average Product is at its peak so that at this level of fertiliser usage MP and AP are equal. Obviously the line drawn to the TP curve from the origin (which gives AP) when it is tangential to the TP curve has the same slope as the TP curve (which gives MP). With higher levels of fertiliser than are shown on the graph, TP and AP can be expected to decline further until they again meet at zero yield somewhere beyond the right-hand edge of the Figure.

Figure 5.2 is a generalised case of the relationship between quantities of one variable input and the quantity of product containing both increasing and diminishing returns and negative Marginal Products. In reality, factor-product relationships could be found from which increasing returns and/or negative Marginal Products were absent, implying differently-shaped TP curves, but for illustrative purposes it is convenient to use a model containing all these features.

What is the Most Profitable Level of Fertiliser to Use?

The most profitable level of fertiliser (in other words, optimising the factor-product relationship) can be worked out from a diagram similar to Fig. 5.2. However, to do this it is necessary to know (a) the price the producer can get for a unit of his product (in our case, barley), and (b) the cost per unit of his factor of production, fertiliser. We also initially assume that the producer operates under Perfect Competition, not an unreasonable assumption in the case of farming, so that however much barley he produces, he will always get the same price per unit and, however much fertiliser he buys, he will have to pay the same price per unit. It is possible to incorporate complications concerning falling product prices with greater outputs and discounts for bulk purchasing etc., but we will avoid them at present.

Product and Value Product

If the price of the product, barley, is known, then instead of expressing

FIG. 5.3   *The Values of Barley Achieved with a Range of Fertiliser*
Usages on a Fixed Quantity of Other Factors of Production

Units of fertiliser	Value of Barley produced (£) (Total Value Product)	Marginal Value Product MVP (£)	Average Value Product AVP (£)
0	0	—	0
1	6	6	6
2	18	12	9
3	40	22	13
4	76	36	19
5	106	30	21
6	130	24	22
7	150	20	21
8	160	10	20
9	166	6	18
10	170	4	17
11	172	2	16
12	170	−2	14
13	166	−4	13
14	160	−6	11
15	152	−8	10

the barley produced by different levels of fertiliser in physical units, we can, by multiplying by the market price, express the product in terms of its *value*.*

Product in physical units × price obtained per unit = Value Product

If, for convenience, we assume that the price barley fetches is £1 per unit, then the figures appearing in Fig. 5.1 for TP, MP and AP at different levels of input (fertiliser) use can be simply re-labelled Total Value Product (TVP), Marginal Value Product (MVP) and Average Value Product (AVP). These are reproduced in Fig. 5.3.

From Fig. 5.3 it can be seen that, for example, when 9 units of fertiliser are used, £166 worth of barley is produced (TVP); the last (9th) unit of fertiliser produces £6-worth of barley (MVP) and the average value of barley produced per unit of fertiliser (AVP) is £18 at this level.

Figure 5.4 shows the TVP, MVP and AVP in graphical form. The curves are identical to Fig. 5.2 except that the vertical axis is labelled in

---

* Some texts prefer to use the term "revenue". Total Value Product becomes Total Revenue Product and so on.

Fig. 5.4 *Cost of a Variable Factor and Value of the Product at Different Levels of Input of the Variable Factor*

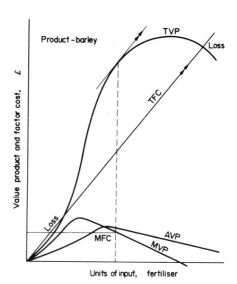

money terms (Value Product) and TP changed to TVP etc. The vertical axis is also labelled "costs" and Fig. 5.4 contains two lines, TFC and MFC, which were not in Fig. 5.2. TFC (Total Factor Cost)* is the total cost of the variable factor, fertiliser, which we are considering at every level of fertiliser use. It is calculated at each level of fertiliser by

Number of units of fertiliser × cost per unit of fertiliser = TFC

Because cost per unit of fertiliser is constant, TFC is a straight line. MFC (Marginal Factor Cost) is the cost to the producer of his last unit of fertiliser. Because the cost of the 4th unit of fertiliser is the same as the cost of the 10th or any other unit of fertiliser, i.e. MFC is constant for all levels of fertiliser use, the MFC curve is a straight line parallel to the horizontal axis.

---

* Note: TFC is often used as an abbreviated form not only of Total Factor Cost, as here, but also of Total Fixed Cost. Care should be taken to distinguish which is implied (see p. 152 *et seq.*).

The Most Profitable Level of Input

(a) The "Total" Approach

The biggest profit will be made when the total value of what is produced (TVP) exceeds by the largest difference the total cost of the factor required to produce it (TFC). In Fig. 5.4 there are two areas of fertiliser input, labelled LOSS, where TFC (the cost of fertiliser) *exceeds* TVP (the value of barley produced). Obviously no rational farmer would use such levels of fertiliser. There is, however, a range of fertiliser usage between these two LOSS areas where TVP is greater than TFC, i.e. a profit will be made. The farmer has to decide where in this range is the *most* profitable level of fertiliser to use.

As profit is the difference between TVP and TFC, the greatest profit will be made where their difference is greatest. On the graph in Fig. 5.4 this means where the *vertical* distance between the two curves is greatest.* This corresponds to the optimum, or most profitable, level of fertiliser to use. Note that the most profitable level of production is not the level at which TVP is greatest, i.e. where the value of barley produced is highest and where yields are greatest. This is because, to achieve these high yields, disproportionately large (and hence costly) quantities of fertiliser are needed because of diminishing returns.

(b) The "Marginal" Approach

The most profitable level of input to use can also be calculated using the cost of additional units of fertiliser (MFC) and the increase in the total value of the product (MVP) which is produced by this additional input. Where the value of the extra yield of grain (MVP) is greater than the cost of the last unit of fertiliser (MFC) used to produce it, the farmer will make a profit on that grain. If he is aiming at maximum profit he will consider using more and more fertiliser until the point is reached where the value of the extra yield of grain (MVP) has dropped because of diminishing returns to a level where it only just offsets the

---

* The point where the vertical distance between TVP and TFC is greatest can be found by drawing a line parallel to TFC and tangential to TVP. The slopes of the TVP and TFC curves will be the same.

cost of the last unit of fertiliser used to produce it (MFC). At this point MFC = MVP. Maximum profit is made at this point. Note that at this point MVP is falling and is below AVP. If more fertiliser were to be used the value of the extra grain (MVP) would be less than the cost of the unit of fertiliser used to produce it, and the farmer's total profit would start to fall.

From Fig. 5.4 it can be seen that the level of fertiliser input where MFC = MVP coincides with the level which was decided as the most profitable (optimum) when TFC and TVP were considered. The two approaches give the same result. In real-world situations the marginal approach is often the more applicable—farmers are concerned with the returns they can expect from using *extra* fertiliser, giving *extra* food to cows etc. in pursuit of higher profits.*

Changes in Prices and Costs

The optimum quantity of a variable input will depend on (a) the price of the factor; (b) the price of the product; and (c) the nature of the Production Function (TP curve). What is an optimal quantity of input under one set of conditions may well not be optimal if prices change, or the Production Function shifts.

If the price of the variable factor (fertiliser in our example) fell, the slope of the TFC line in Fig. 5.4 would become less steep and the MFC line would fall. This would result in the MFC and MVP curves intersecting at a point corresponding to higher levels of fertiliser input, and the maximum vertical distance between TFC and TVP would also be shifted further to the right. This is shown in Fig. 5.5 where it can be seen that, to achieve maximum profit, when the price of the factor drops, more is used (an increase from OF_1 to OF_2).

The effect of a rise in the price of the product is to alter the level of all the value product curves in Fig. 5.4. It is as if the curves were all stretched upwards. (This is shown in Fig. 5.6.) A similar effect is caused when a new higher yielding variety of barley is introduced which yields

---

* MVP is the same as the slope of the TVP curve. Similarly, MFC is the slope of the TFC. Hence, if the greatest level of profit indicated by the maximum vertical difference between MVP and TFC is when their slopes are the same, then MVP must equal MFC at this maximum profit position.

greater weights of grain than the old at the same levels of fertiliser use. The greatest profit level now occurs at a higher level of fertiliser usage.

How Many Optima are There?

When the price of the factor or the product changes, the optimum usage of fertiliser changes too. For every cost/price situation there will be an optimum quantity of variable factor to use. Furthermore, in agriculture the weather has the effect of lowering the Production Function in a poor year and raising it in a good year and this will also shift the optimum quantity of fertiliser. Before deciding on the optimum quantity of an input, a farmer—or entrepreneur—has to weigh in his mind not only the present cost of the input but also what price he will get for his product and the likelihood of shifts in his production function. It is small wonder that, in view of the many unknowns, guesses and rules of thumb are used by farmers in place of precise estimates.

Fig. 5.5   *The Optimum Factor—Product Combination,*
*at Different Levels of Factor Cost**

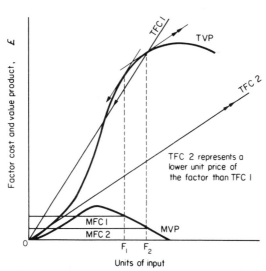

* The value product curves can be read as physical output
curves simply by dividing by the price of the product

FIG. 5.6 *The Optimum Factor-Product Combination at Different Levels of Product Price*

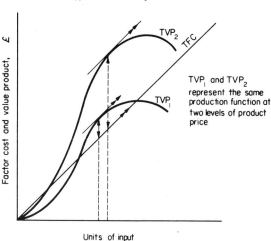

TVP₁ and TVP₂ represent the same production function at two levels of product price

An Alternative Approach to Optimising the Factor-Product Relationship

So far we have been answering the question "How much of one variable factor (input) should I use to generate most profit, all other factors being fixed?" An alternative approach is to ask "What level of *output* of a product should I aim for to make most profit?" Situations in agriculture in which such questions might be asked would be in deciding the best yields to aim for with animals or crops.

For our example we will take the problem of deciding on the most profitable level of output of eggs from poultry from the range of performance which can be arrived at by altering the feeding level. Note that all other factors of production (labour, housing etc.) are assumed to be held constant—only the quantity of food is varied to give the different levels of output per bird. Consequently, food cost is the only variable cost and all other costs are assumed fixed; they do not vary according to the quantity of eggs produced. For the time being we will ignore these fixed costs and simply attempt to answer the question: "Given that we already have a fixed quantity of buildings and equipment, fully stocked with birds etc., with food being the only factor which we can vary, and given that we know the price of eggs and the cost of food, what is the most profitable level of performance to aim for?"

The costs of the food required to produce a range of egg-laying performances are shown in Fig. 5.7. In addition to the total food cost column (which should also be labelled Total Variable Cost, or TVC, because food is here the only variable input) there are columns for the average food cost per egg (which similarly could be labelled Average Variable Cost, AVC) and the Marginal Cost of production. While the derivation of the average food cost column is obvious, being the total food cost divided by the number of eggs produced, the Marginal Cost figures require an explanation.

MARGINAL COST (MC) OF PRODUCTION IS THE ADDITION TO TOTAL COSTS CAUSED BY PRODUCING THE LAST (OR MARGINAL) UNIT OF *OUTPUT*.

Note that MC falls first with increasing output and then increases. This is closely linked with increasing and diminishing returns to the input, food, discussed earlier. MC eventually rises because perhaps the birds are not genetically capable of assimilating larger quantities of food efficiently and wastage occurs. Like MC, AVC also falls and then

FIG. 5.7   *Costs of Food in the Production of Eggs at Different Levels of Output per Hen (All other inputs are assumed to be constant)*

Output Eggs/hen/yr	Total Food Cost (TVC) p	Marginal Cost p	Average Food Cost (AVC) p
49	80		
50	81	1	1.6
99	109.5		
100	110	0.5	1.1
149	149		
150	150	1	1.0
199	250		
200	253	3	1.25
249	420		
250	427	7	1.7
299	700		
300	713	13	2.3

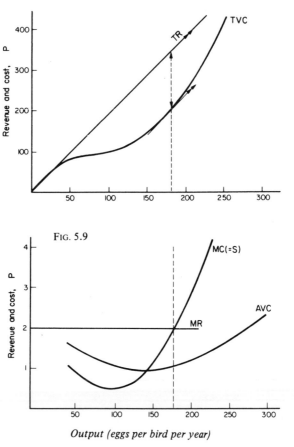

FIG. 5.8   *Costs of Production of Eggs and*
Optimum Level of Output
Food is assumed to be the only variable input

Output (eggs per bird per year)

rises although the point of lowest Average Variable Cost is not at the same output as the lowest Marginal Cost. This is easily seen when the figures are plotted (Figs. 5.8 and 5.9). Note that, while we have hitherto placed units of *input* on the horizontal axis of graphs relating input to output, we now reverse the labels and units of output appear on the horizontal axis.

The Optimum Level of Output

To calculate the most profitable level of egg-production two further lines must be drawn relating to Total Revenue and Marginal Revenue. On the TVC curve the Total Revenue line must be drawn—this shows the revenue which is brought in by the various levels of egg production and is calculated as follows:

Total Revenue = Number of eggs produced × price received for eggs

It is shown in Fig. 5.8. It is a straight line because we assume that eggs are sold on a perfectly competitive market. The most profitable level of production is where TR exceeds TVC (total food cost in this example) by the greatest vertical distance.*

MARGINAL REVENUE (MR) IS THE INCREASE IN TOTAL REVENUE CAUSED WHEN OUTPUT IS INCREASED BY ONE UNIT; i.e. it is the extra revenue brought in by the last egg produced. Under conditions of perfect competition the producer will get the same price for his eggs irrespective of how many he sells, so the MR of the 4th egg will be the same as the MR of the 400th egg. MR is constant with different levels of output.

MR has been superimposed on the MC curve in Fig. 5.9. It is obviously in the interest of a farmer's profit for him to keep increasing his output of eggs by feeding his birds more as long as the extra revenue produced is greater than the extra cost incurred, i.e. while MR is greater than MC. Profit will be *reduced* if production is expanded beyond where MR = MC because the additional costs exceed the extra revenue. Maximum profit is hence made at the output where MC = MR. This corresponds to the same optimum output shown in Fig. 5.8.

Price and Cost Changes

As has been shown earlier, the optimum solution of the factor-product problem is only optimal for a particular set of price levels. A fall in product price will shift the TR and MR lines. The MR line will fall, intersecting the MC line at a lower level of output. If the product price rose, MR would rise and intersect MC at a higher level of output.

---

* This point can be found by drawing a tangent to the TVC curve parallel with the TVR line.

Hence, for a firm aiming to maximise its profits in perfect competition, its MC curve is the same as its supply curve because the MC curve shows the quantities it is willing to produce (and hence supply to the market) at different levels of product price.*

Changes in the price of the variable input, food in our example, will shift the TVC, AVC and MC curves and alter the optimal level of output accordingly.

Reconciling the Methods of Optimising the Factor-Product Relationship

Two ways of optimising the factor-product relationship have been examined. The first attempted to find the optimum level of one variable input. The second attempted to find the optimum level of *output*. If the same one product—one variable input example had been used all through, it would have been found that the results were perfectly compatible—by using the optimum quantity of fertiliser on barley found in the first approach, we should have produced that quantity of barley which we would have shown to the optimum quantity to produce by the second approach. In producing the optimum quantity of eggs, our second example, we should have used the optimum quantity of food. This is because both approaches to each problem are based on the same production function and the same prices of inputs and outputs.

Optimising the Factor-Factor Relationship

Up to this point we have been looking at the most profitable level of one variable input (or factor of production), with all other factors of production held constant, or "fixed". Optimising the factor-factor relationship involves choosing the most profitable combination of inputs *when more than one input is variable.* For example, if a cow can be fed on different combinations of hay and barley to yield, say 1,000 gallons per year, optimising the factor-factor relationship involves

---

* This point was also made in Chapter 4 when referring to Fig. 4.2. The MC curve is only the supply curve when price is greater than Average Variable Cost (see later) because no production will occur unless these costs are covered. In this context price must be greater than the lowest point on the Average Cost curve of feed before any production will occur.

FIG. 5.10 *Yields of Milk Produced from*
Combinations of Two Variable Inputs,
Hay and Barley

Tons of barley/year	Gallons per year					
6	900	(1000)	1075	1100	1110	1115
5	825	975	1050	1090	1106	1112
4	700	900	(1000)	1075	1100	1110
3	500	750	900	(1000)	1075	1100
2	100	400	500	750	900	(1000)
1				400	500	750
	2	4	6	8	10	12

Tons of hay/year

devising the cheapest (or least cost) combination of the two feeds.

Fig. 5.10 shows the quantities of milk that can be produced from a cow by different combinations of quantities of hay and barley, all other factors of production being held constant.* Notice that in this example each factor is subject to Diminishing Returns; for instance, choose a level of hay usage (say 4 tons) and go up the column of milk yields produced by 1, 2, 3, 4 tons of barley. The increase in yield produced by each extra ton of barley progressively diminishes. The same Diminishing Returns can be seen if a level of barley usage is selected and the increase in yield with increasing hay usage is examined. The figures in this table are presented in graphical form in Fig. 5.11 in which a line has been drawn connecting all those combinations of hay and barley which produce 1,000 gallons of milk per year. It is called the 1,000 gal. Isoquant. Other lines have been drawn corresponding with yields of 900 gals., 1,100 gals. etc.

---

* The Production Function describing the relationship between milk yield and hay and barley usage might be written in its simplest, symbolic form as follows:

  Milk yield = f (hay usage, barley usage, all other inputs held constant)

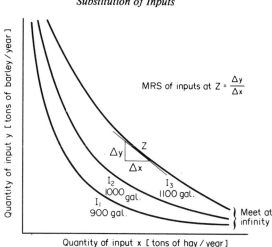

FIG. 5.11   *Marginal Rate of Substitution of Inputs*

AN ISOQUANT IS A LINE CONNECTING THE VARIOUS COMBINATIONS OF TWO INPUTS WHICH PRODUCE THE SAME LEVEL OF OUTPUT.

Isoquants have a number of properties:

(i) They are normally convex to the origin. Hence isoquants are normally steep near the vertical axis and less steep further away. In this example, situations where all barley and no hay is used or vice versa are not technically possible, so the isoquant never touches the axis (see Fig. 5.10).

(ii) The number of units of barley which must be given up to keep yield the same if hay usage is stepped up by one unit is the Marginal Rate of Substitution (MRS) of Inputs.*

$$\text{MRS} = \frac{\text{change in number of units of input on vertical axis required to keep output constant}}{\text{change in one unit of input on horizontal axis}}$$

---

* It is of course the same as the number of *extra* units of barley which must be fed to maintain yield if one less unit of hay is fed. The MRS between two inputs is also sometimes called the Rate of Technical Substitution.

$$= \frac{\Delta \text{ input } y}{\Delta \text{ input } x} \text{ at a constant level of output}$$

The MRS changes as we move along the horizontal axis away from the origin. When little hay but a lot of barley is fed, using one extra unit of hay will result in a large reduction in barley usage for a given level of milk yield, but if the balance is already heavily weighted towards hay, using one further unit will result in little barley saving. The MRS at any point is numerically the same as the slope of the isoquant (see Fig. 5.11); it should be accompanied by a negative sign since the change in quantity of one of the inputs is negative.

(iii) As long as each input and the product is easily divisible into small units, a new isoquant can be drawn between any two existing isoquants. Isoquants further from the origin imply higher levels of output.

Notice that these characteristics of isoquants are very similar to those of the Indifference Curves encountered in the Theory of Consumer Choice, Chapter 2; for an individual, a curve corresponded to a particular level of satisfaction given by combinations of two goods, whereas the isoquant corresponds to a particular level of output for a firm.

Fig. 5.12 shows the straight-line isoquant of two inputs which are perfect substitutes for each other. Apparently oil and alcohol are perfect substitutes in the production of synthetic rubber and, at one level of

FIG. 5.12 *Isoquants where Inputs are Perfect Substitutes*

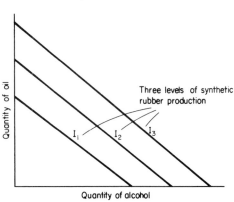

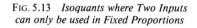

FIG. 5.13   *Isoquants where Two Inputs can only be used in Fixed Proportions*

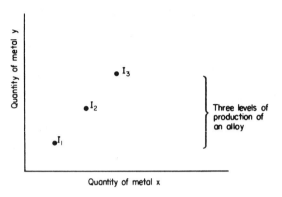

output, one unit of one can always be replaced by the same number of units of the other irrespective of the combination in use.

Fig. 5.13 shows the opposite case to Fig. 5.12. Only one proportion of the two inputs can be used to produce a given level of output, and so the isoquant takes the form of a single point. An example might be the production of a specific alloy from two elemental metals.

FIG. 5.14   *Isoquants and an Iso-cost Line for Milk Production with Hay and Barley as Variable Inputs*

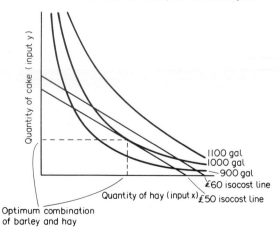

Cost and the Optimum Combination of Inputs

Fig. 5.14 is the same as Fig. 5.11 but with two additional lines—iso-cost lines labelled "£50" and "£60". The £50 iso-cost line joins all the combinations of hay and barley, the two inputs, which can be bought for the same £50 outlay by the producer. (It is equivalent to the Budget Line encountered in the Theory of Consumer Choice.) Similarly the £60 iso-cost line joins all those combinations which cost the producer £60.

AN ISO-COST LINE JOINS THE VARIOUS COMBINATIONS OF QUANTITIES OF TWO VARIABLE INPUTS WHICH CAN BE PURCHASED FOR THE SAME TOTAL COST.

Note that there are two combinations of hay and barley which can be bought for £60 which will produce 1,000 gallons (where the £60 iso-cost line and 1,000 gal. isoquant cut), but only one combination that will allow 1,000 gallons to be produced for £50. This is where the £50 iso-cost line and the 1,000 gal. isoquant are tangential. This is the optimum combination of hay and barley for the producer to use to produce 1,000 gallons, since it is the least-cost way of achieving that output.

At the point of tangency the slopes of the isoquant and the iso-cost line are the same. It has already been established that the slope of the isoquant at any point equals the MRS at that point.

$$\text{Slope of the isoquant} = \text{MRS} = \frac{\Delta \text{ input y}}{\Delta \text{ input x}} \quad \text{(see Fig. 5.11 if you are not clear on the estimation of MRS)}$$

$$\text{Slope of the iso-cost line} = -\frac{\text{quantity of y bought for £50}}{\text{quantity of x bought for £50}} = -\frac{\text{price of x}}{\text{price of y}}$$

Therefore, at the optimum combination of inputs

$$\text{MRS} = \frac{\Delta \text{ input y}}{\Delta \text{ input x}} = -\frac{\text{price of x}}{\text{price of y}}$$

It can be shown that, when implying infinitely small changes in x and y:

AT THE OPTIMUM COMBINATION OF TWO VARIABLE INPUTS THE SAVING IN COST RESULTING FROM REDUCING THE QUANTITY OF ONE INPUT BY AN INFINITELY SMALL AMOUNT WILL BE EXACTLY OFFSET BY THE EXTRA COST OF USING MORE OF THE OTHER FACTOR NECESSARY TO MAINTAIN THE LEVEL OF PRODUCTION.

This statement can be extended to any number of variable factors.

FIG. 5.15 *A Change in the Relative Prices of Inputs*

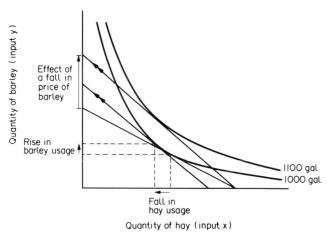

Quantity of hay (input x)

When implying *more than* infinitely small changes in x and y the fall in costs resulting from using less of one input will be MORE THAN OFFSET BY the extra cost of the other input, i.e. the combined costs of both inputs to produce a given level of output will rise.*

Changes in the Price of Inputs: Price/ Substitution Effect

Returning to the hay and barley example, if the price of one input, say barley, fell, a greater quantity could be purchased for £50. Thus the slope

---

* It can also be shown that at the optimum combination of two variable inputs, the ratio of their prices will equal the ratio of their Marginal Products, i.e. the addition to total output resulting from the last unit of each input.

$$\frac{\text{Marginal Product of x}}{\text{Marginal Product of y}} = \frac{\text{price of x}}{\text{price of y}}$$

In rearranged form this becomes

$$\frac{\text{MPx}}{\text{price of x}} = \frac{\text{MPy}}{\text{price of y}}$$

This can be extended to greater numbers of variable inputs; at the least-cost combination the addition to total output resulting from the marginal unit of each variable input will be proportional to each input's price.

of the iso-cost line would shift, shown in Fig. 5.15. For £50 the producer can now produce 1,100 gallons. If the farmer still wished to produce only 1,000 gallons, the new least-cost diet is found by shifting the iso-cost line, keeping its slope constant, until it is tangential to the 1,000 gallon isoquant. Note that in this new least-cost diet, barley will have been substituted for hay, with less hay but more barley in the optimum mix than under the previous set of prices.

FIG. 5.17   *The Effect of a Change in Input Price where*
Two Inputs can only be Combined in Fixed Proportions

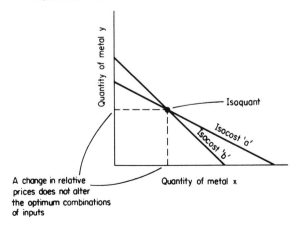

Fig. 5.16 shows the effect that a change in relative prices can have on the choice of inputs where straight line isoquants are involved. In the first situation (iso-cost line "a") all of input y is used and none of x. When relative prices change (producing iso-cost line "b") no y and all x is used—a small change in relative price will cause a large change in the optimum choice of inputs. Only when the iso-cost line and isoquant are of identical slopes will a combination of x and y inputs be "optimal", and under that situation *any* combination on the isoquant is optimal because all combinations cost the same.

Fig. 5.17 shows the case of a single-point isoquant (the hypothetical case of an alloy quoted earlier). Here changes in the relative prices of inputs have no effect on the least-cost combination of inputs to produce a given level of output. A less extreme form of insensitivity to changes in the relative prices of inputs is encountered with "lumpy" inputs, i.e. those which can only be had in relatively large indivisible units such as tractors. Fig. 5.18 shows the case of the quantity of labour (man-days) and tractors required to produce a given quantity of hedge cutting. The isoquant is a series of discrete points corresponding to whole numbers of tractors. Because of the "corners" on the isoquant formed by this series

Fig. 5.18   *The Effect of Changing Input Prices*
where Inputs are "lumpy"

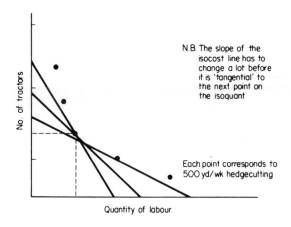

N.B. The slope of the isocost line has to change a lot before it is 'tangential' to the next point on the isoquant

No. of tractors

Each point corresponds to 500 yd/wk hedgecutting

Quantity of labour

FIG. 5.19   *The Optimum Combination of
Two Variable Inputs at Different Levels
of Output*

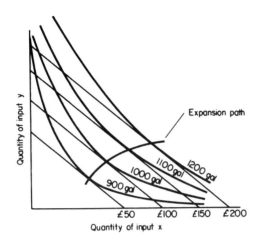

of points, the cost of labour can change considerably, changing the
slope of the iso-cost line, before the least-cost combination is altered.*

The Effect of Expanding the Scale of Operation

Fig. 5.19 shows isoquants for a range of milk outputs and iso-cost
lines enabling the least-cost combination for each output to be found.
A line connecting these optima for various levels of output is called the
EXPANSION PATH. Note that the inputs are not necessarily in the
same relative proportions at different levels of output. Obviously the
prudent entrepreneur wishes to operate at the best scale of production,
and how far he proceeds along the Expansion Path is determined by the
revenue brought in by each level of output, i.e. the revenue corresponding
to each isoquant, and the level of cost, represented by the iso-cost line,
necessary to achieve it. The profit maximising entrepreneur must aim
for that output which gives him the biggest difference between revenue

---

* This is admittedly a superficial treatment of the "lumpy input" problem and ignores
such questions as the possibility of only partly using the capacity of the machine to
replace labour. However, no mention at all would be a greater error.

and costs. Knowing the optimum combination of variable inputs for each level of output, and hence the lowest *combined* cost for each level, he can decide the optimum level of output using the Marginal Cost and Marginal Revenue approach described in the section on the Factor-Product Relationship.

Optimising the Product-Product Relationship

"Optimising the product-product relationship" involves the entrepreneur allocating the limited resources at his disposal in order that they may bring in the greatest revenue. For example, a farmer with a farm of a given size, a fixed labour force, stock of machinery and buildings and a fixed amount of working capital has to choose to allocate them between, say, potato production and sugar beet production; he can produce all of either commodity or some of each. Fig. 5.20 shows the various combinations of potatoes and sugar beet that can be produced in a given period using the same bundle of resources—those which exist on the farm in question. This curve is therefore called an ISO-RESOURCE CURVE because it relates to the same lot of resources (land, labour, capital, entrepreneurship). It is also called the Production Possibility Boundary as, while any combination of two products in question lying inside or on this curve can be produced with the existing resources and technical knowledge, combinations lying further out from the origin are not possible. We need not concern ourselves with combinations lying inside the curve, only those right on the boundary, since we can assume that the profit-maximising farmer will aim to keep his productive resources fully utilised.

Potatoes and sugar beet are COMPETITIVE PRODUCTS because if more of one is produced *less* of the other can be produced. The slope of the iso-resource curve is negative and it is numerically the same as the Marginal Rate of Substitution of Products;* the MRS between competitive products has a negative coefficient since a positive change in the amount of one product produced must be accompanied by a negative change in the amount of the other which can be produced.

---

* Also called the Marginal Rate of Transformation between products, or the Rate of Technical Transformation.

FIG. 5.20 *The Combinations of Two Products that can be Obtained from a Given Bundle of Resources*

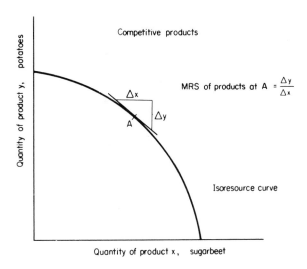

Marginal Rate of Substitution of Products	=	Number of units of product y that have to be foregone for an increase in production of 1 unit of product x
	=	$\dfrac{\Delta y}{\Delta x}$

Fig. 5.21 shows that the slope of the iso-resource curve is not always negative. If a wholly cereal farm devotes part of its acreage to growing grass (and producing hay) its total cereal production may *increase* because of higher yields per acre. This is caused by the grass "break" enabling cereal diseases to be held in check, soil structure to be improved etc. Over the range AB in Fig. 5.21 hay and wheat are COMPLEMENTARY PRODUCTS since producing *more* hay enables *more* wheat to be produced. (MRS is positive over this range.) However, beyond B the extra cereal yield per acre is insufficient to offset the loss of production caused by switching acres from cereals to grass, and hay and cereals become competitive products, the MRS between them taking a negative sign.

Fig. 5.21   *Iso-resource Curve of
Complementary Products*

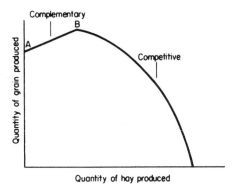

Fig. 5.22   *The Optimum Combination of Two
Products from a Given Bundle of Resources*

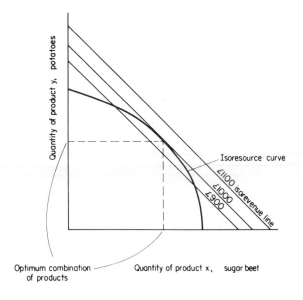

Revenue and the Optimum Combination of Products

Not all combinations of products will bring in the same amount of money to the entrepreneur. The best combination of products (or enterprises) to be produced from a given bundle of resources is that which brings in the greatest revenue. Fig. 5.22 is the same as Fig. 5.20, except for the introduction of iso-revenue lines labelled £900, £1,000 and £1,100.

AN ISO-REVENUE LINE CONNECTS THE COMBINATIONS OF QUANTITIES OF TWO PRODUCTS WHICH BRING IN THE SAME REVENUE.

An iso-revenue line can be constructed by joining the quantity of the good on one axis which could be sold for, say, £1,000 with the quantity of the other good on the other axis which could be sold for £1,000. Fig. 5.23 shows that two combinations of potatoes and sugar beet could be produced to bring in £900, but only one combination to bring in £1,000. This is the highest revenue attainable and hence this combination of products is the best (or is optimal).

At the optimal point

Slope of iso-resource curve = Slope of iso-revenue line (Note: both are negative)

Slope of iso-curve = MRS $= \dfrac{\Delta \text{ product } y}{\Delta \text{ product } x}$ with the same bundle of resources

Slope of iso-revenue line $= -\dfrac{\text{quantity of y which will bring in £n}}{\text{quantity of x which will bring in £n}}$

At the optimum combination of products x and y

$$MRS = \frac{\Delta \text{ product } y}{\Delta \text{ product } x} = \frac{- \text{ Price of } x}{\text{Price of } y}$$

It can be shown that, when considering infinitely small changes in x and y:

AT THE OPTIMUM COMBINATION OF TWO PRODUCTS FROM A GIVEN BUNDLE OF RESOURCES ANY RISE IN REVENUE CAUSED BY PRODUCING AN INFINITELY SMALL AMOUNT MORE OF ONE PRODUCT WILL BE EXACTLY OFFSET BY THE LOSS RESULTING FROM THE REDUCTION IN PRODUCTION OF THE OTHER PRODUCT.

When considering greater-than-infinitely small changes in x and y, the rise in revenue caused by producing more of one product will be

FIG. 5.23   *The Effect on the Optimum*
Combination of Products of a Change
in Relative Product Prices
(non-constant MRS of products)

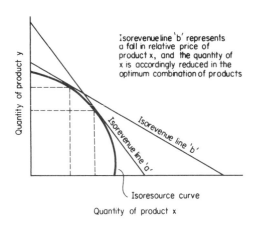

Isorevenue line 'b' represents
a fall in relative price of
product x, and the quantity of
x is accordingly reduced in the
optimum combination of products

Isorevenue line 'b'

Isorevenue line 'a'

Isoresource curve

Quantity of product x

MORE THAN OFFSET by the loss resulting from the reduction in
production of the other product. This can be extended to more than
two products and this optimum allocation of a given quantity of
productive factors between a range of possible uses is yet another
manifestation of the Principle of Equimarginal Returns encountered
already in Chapters 2 and 4.

Changes in Relative Prices of Products

A change in the relative prices of products will cause the slope of the
iso-revenue line to change (Fig. 5.23) producing a new optimum combination of product quantities. If the price of one product rises, it will be
favoured in the new combination.

Straight Line Iso-Resource Curves

Where there are no Diminishing Returns the iso-resource curve will
be a straight line. For example, if a factory could easily switch from

producing plastic dustbins to producing plastic buckets with no Diminishing Returns small changes in relative price would cause a complete swing from one good to the other (Fig. 5.24).

Complementary Products

Because of the "bump" on the iso-resource curves for these products, large changes in relative prices may be required before large changes in the optimum combination of products will occur (Fig. 5.25). With the complementary products example cited above (grass and cereals), a fairly heavy loss on the grass crop viewed alone would be required before no grass would be grown because of the boost it gives to cereal production.

Time and Scale of Production

The three important relationships (factor-product, factor-factor, product-product) have been examined in some detail. It is now necessary to return to a comment that was made at the start of this section on Production Economics—the importance of TIME.

FIG. 5.24 *The Effect on the Optimum Combination of Products of a Change in Relative Product Prices where MRS of Products is Constant*

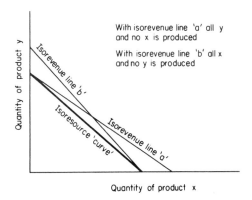

With isorevenue line 'a' all y and no x is produced

With isorevenue line 'b' all x and no y is produced

Quantity of product y

Isorevenue line 'b'

Isoresource 'curve'

Isorevenue line 'a'

Quantity of product x

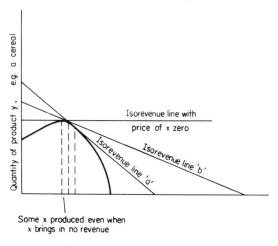

FIG. 5.25   *The Effect on the Optimum Combination*
of Products of a Change in Relative Prices
of the Products with the Products
Complementary over a Range

Fixed Costs and Variable Costs

The fact that some costs can be considered fixed while others are variable has already been mentioned in a variety of contexts. A *variable cost* of production is one that varies with the level of output. A *fixed cost* is one which does not vary with output. Which costs are fixed and which variable depends on the time period being considered. If a farmer has rented a farm and hired a labour force, he is committed to pay the rent and wages whether he decides to produce one ton of cereals per acre or two tons. Conversely, seed costs are variable costs when viewed before planting—they will depend on the quantity of grain the farmer decides to produce. In the long term *all* costs are variable—the farmer could increase or reduce the size of his labour force or the acreage of his farm. On the other hand, the shorter the time period the more costs become "fixed". A farmer, an hour before harvesting his cereals, cannot change anything to increase their yield except the efficiency of the harvesting process—he is committed to all the costs of production except those of harvesting. Given more time he could have varied fertiliser use, herbicide use, seed rates etc., but in the short run these are "fixed".

The table in Fig. 5.26 shows the fixed and variable costs faced by the operator of a brewery at different levels of output. The fixed costs, which do not vary with the number of barrels of beer produced per month, consist of the lease of the factory, the monthly wage bill, depreciation of equipment etc. The variable costs consist of barley costs, malt costs, hop costs, colouring and water costs.

Note that (a) because increases in output spread fixed costs over more barrels of beer, Average Fixed Costs (AFC) per barrel fall consistently.

(b) Total Variable Costs rise with output but not in a fixed proportion, so that Average Variable Costs (AVC) per barrel first fall, then rise. AVC initially decline because inputs are being combined

Fig. 5.26   *Fixed, Variable and Total Costs of Production for Beer Manufacture*

Output (barrels per month)	Fixed Costs Total	Fixed Costs Average per barrel	Variable Costs Total	Variable Costs Average per barrel	Total Costs Total	Total Costs Average per barrel	Marginal cost of producing the last barrel*
100	240	2.40	250	2.50	490	4.90	4.90
200	240	1.20	450	2.25	690	3.45	2.00
300	240	0.80	615	2.05	855	2.85	1.65
400	240	0.60	800	2.00	1040	2.60	1.85
500	240	0.48	1010	2.02	1250	2.50	1.50
600	240	0.40	1290	2.10	1530	2.53	2.70
700	240	0.34	1580	2.26	1820	2.60	3.20
800	240	0.30	2000	2.50	2240	2.80	4.20

* MC of the nth barrel of beer is calculated by Total Cost of 'n' barrels minus the Total Cost of 'n − 1' barrels. Because Fixed Costs are constant with output, MC is also the same as the increase in Variable Costs, i.e. VC of 'n' barrels minus VC of 'n − 1' barrels. The outputs shown (100, 200 etc.) do not contain the information from which the MC of (say) the 300th barrel has been calculated, but the following figures serve as an example of how MC of the 300th barrel is derived.

Output	Fixed Costs	Variable Costs	Total Costs	Marginal Costs
299	240.00	613.35	853.35	
300	240.00	615.00	855.00	1.65

FIG. 5.27 *Fixed, Variable and Total*
Costs of Beer Production

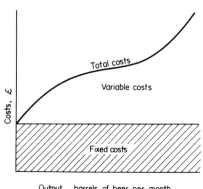

Output, barrels of beer per month

more efficiently, but they eventually rise because of inefficiencies arising from such factors as the overwork of the equipment, over-stretching of the labour force etc.

(c) Average Total Costs (ATC) per barrel is a combination of fixed plus variable costs. At first the fall in Average Variable Costs with output reinforces the fall in Average Fixed Costs, enabling Average Total Costs to fall. However, the eventual rise in Average Variable Costs more than offsets any fall in Average Fixed Costs, so that Average Total Costs rise.

(d) the Marginal Cost (MC) i.e. the addition to total cost attributable to the last barrel, first falls and then rises.

Costs from Fig. 5.26 are plotted in Figs. 5.27 and 5.28; Fig. 5.27 shows how total cost varies with output, and Fig. 5.28 shows AFC, AVC, ATC and MC. Note that the rising MC passes through the lowest point of the AVC and ATC curves.

What is the Lowest Price at Which Beer will be Produced?

The beer producer will equate MC with MR to maximise profits. Remember that in perfect competition MR = AR (see Chapter 4). To continue in production the entrepreneur must receive a price (AR) for

Fig. 5.28   *Average and Marginal Costs of Beer Production*

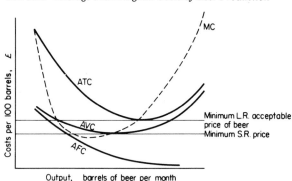

his beer so that both his variable and fixed costs are covered. These are lowest at the lowest point of the ATC curve—he must get at least this price to continue in production in the long run. However, in the short term he will continue to produce as long as his variable costs are covered, so the minimum short-run price he can accept is the lowest point on the AVC curve. In other words, although the brewer has to have a price of £2.50 per barrel to permanently stay in business, he will not stop production if prices slump temporarily unless they fall below £2.00 per barrel because any price above £2.00 covers his variable costs and

Fig. 5.29   *Costs of Production at Three Scales of Milk Production (hypothetical)*

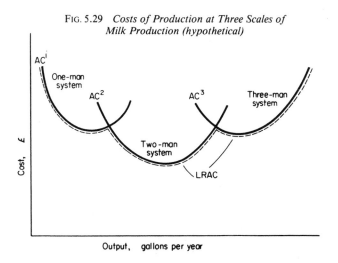

leaves a small margin to help pay the fixed costs which he has to bear whether he produces much or little beer (see Fig. 5.28).

Economies and Diseconomies of Scale

Fig. 5.29 shows three sets of AC curves* for three different sizes of dairy enterprise. These correspond to three different quantities of fixed factors of production; the smallest size corresponds to an enterprise based on a small parlour designed to be operated by one man, the medium-size enterprise to a larger parlour operated by two men and the largest to a three-man system. Once a farmer has built a one-man system he is stuck with it—the costs of running it are "fixed" and he cannot readily scrap it and set up a two-man system. These three sets of equipment plus staff represent three possible *scales of production* where some costs are fixed (depreciation and probably labour) whereas others (e.g. feed) are still variable. Although there is some flexibility in the number of cows each system can handle, both the under-utilisation of the fixed equipment and overstretching its designed capacity will result in higher average costs of milk production. Only when a period of, say, more than five years is reviewed can the farmer think of changing the *scale* of his enterprise.** In Fig. 5.29 the line of dashes shows the minimum average cost at which it is possible to produce each level of output, given the time necessary to alter the scale of production, i.e. when all costs can be considered variable—it is termed the Long Run Average Cost line.

In Fig. 5.29 the level of the AC curve is lower for the two-man dairy enterprise than for the others. This means that such a system could produce at a lower cost per gallon than either larger or smaller scales of production. Being larger than a one-man operation has both advantages and disadvantages (which will be discussed below), but on balance the two-man enterprise benefits in terms of lower costs; going larger than

---

* From here on, AC implies that both variable and fixed elements are included (i.e. = ATC).
** Some prefer to use the term Economies of Size rather than Economies of Scale for situations where not all inputs are increased in the same proportion. In the dairy herd example it is implied that one farmer can choose between the three different sizes of herd and so the input he, himself, represents is not increased. This is the case in almost all practical situations.

FIG. 5.30 *Falling and Rising LRAC Curve*

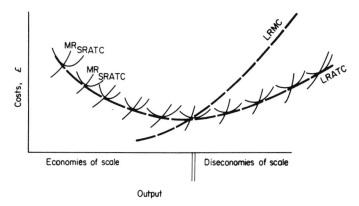

a two-man system incurs disadvantages which more than offset any advantages, so that costs are again higher.

Fig. 5.30 shows another example with a wider range of possible scales; the Long Run Average Cost line (showing the lowest possible average costs of production at each level of output when scale is not "fixed") is much smoother than in Fig. 5.29 because of the greater number of short-run AC curves it envelops. When the LRAC line is falling *economies of scale* are said to operate, and when rising *diseconomies of scale* operate. If over a range the LRAC is horizontal and flat, neither can be said to operate. With the LRAC curve is associated a Long Run MC curve.

Sources of Economies of Scale

Economies of scale arise from a range of sources, some of which are more relevant to farming than others. They are:

    (a) Economies in the use of *land*. Except with some farm crops, it is rare that doubling the output of an enterprise requires double the land area, so that costs per unit of output will be lower than with the larger scale of production.

    (b) Greater efficiency in the use of *labour*.

       (i) Labour can specialise, develop skills and hence produce more.

Specialisation (sometimes called the Division of Labour) only becomes possible with the growth of firms. For example, on a one-man farm the farmer may have to milk cows, drive the tractor and manage the business. On a larger farm specialists in each type of job can be employed who should be better at their individual jobs than the all-rounder and the farmer can concentrate his efforts on managing the business. Considerable savings in time are also made because one man has no longer to keep switching between jobs which may require different clothing or levels of hygiene, or be in different places.

(ii) With higher output, more machinery and more powerful machines can be used. For example, one man can drive a 10 h.p. tractor or a 100 h.p. model, and the cost of his labour can be spread over a greater volume of output with the larger machine.

(c) Economies in the use of *capital*. Larger machines and buildings are usually cheaper in relation to what they can do than smaller ones. For example, a 40-metre length of fencing will enclose a square plot of 100 sq.m. However, an 80-metre length of fence will enclose a square plot of 400 sq.m. so that the cost of fencing *per sq.m. enclosed* falls with increasing size of plot. Similarly, doubling all the dimensions of a square building (including its height) will increase its internal volume eight times. If the same materials could be used, the larger building would have a much lower cost *per unit of volume*. For similar reasons, the price of tractors per horsepower generally falls with increasing size of tractor (at least until the very largest tractors where their small-scale production increases their price). Some units of capital are expensive and indivisible and are only justified by large-scale enterprises—for example, a large farm may be able to fully utilize a machine for laying drains in fields whereas a small farm may have to hire a contractor and machine, at a greater cost per acre. A large business may acquire a computer to work out its accounts and the cost per invoice handled will be lower than that of the clerks used by a smaller firm.

Cost advantages coming from the ability to fully utilise indivisible resources are further illustrated by the better matching of units of equipment. For example, a production process using four machines in succession and an output of 10 units per hour may seriously under-utilise some machines while others are used to full capacity. In the following figures the *indivisibility* of machines A, B and D (it is not possible to have a third of machine A etc.) leads to a situation where larger scale production results in economies.

Machine	A	B	C	D
Annual capacity (units per hour)	10	10	10	10
Potential capacity	30	20	10	40

The bottleneck in production is machine C. The lowest output where all machines could be fully utilized, implying the lowest average cost per unit, is 120 units per hour (the lowest common multiple of 10, 20, 30, 40) when 4 type A machines, 6 type Bs, 12 type Cs, and 3 type D machines would be employed. Thinking of machines not individually but as they fit into the whole process of production is called a "systems" approach.

(d) *Economies of administration.* Increasing the scale of production does not necessarily need a proportional increase in clerical and administrative staff. A clerk can as easily write an order for fifty tons of raw material as for five. Managers can often as easily manage a big business as a smaller one. Also firms which are large can often afford higher salaries for their management and, for a little extra salary, can attract the best managerial talent which more than earns its extra cost.

(e) *Economies of material.* The waste product of a large firm may often be turned profitably into a saleable by-product. A trite example is that manure from a one-horse stable may be difficult to dispose of, but a large racing stable can sell its manure profitably for mushroom compost.

(f) *Economies of marketing.* Large-scale purchasing of inputs often achieves discounts. In selling, large-scale firms are often able to secure an outlet using a contract, thereby enabling them to specialise and thus to lower the cost. For example, a chain store may be willing to take lettuces from a grower on contract at a predetermined price if the grower is large enough to be able to guarantee the quality and quantity he can supply. With commodities where a travelling sales force or advertising is important, larger sales mean that these selling costs are spread over more units.

(g) *Economies of finance.* Large firms can often borrow more easily and at preferential rates. Most farms are small private businesses, unable to finance investment by issuing new shares as could a large public company.

(h) *Economies of research and development.* Large firms may be able to afford R & D departments which can give their products a competitive edge (more relevant to the industries supplying agriculture than to farming itself).

(i) Other economies.* A large firm may be able to provide social and other welfare facilities and so retain a band of loyal, happy and productive workers.

In agriculture the very small farm frequently suffers primarily from its inability to utilise fully the labour and essential equipment found on it. For example, the minimum labour force of one man (the farmer) may not be fully utilised yet, because of his *indivisibility*, it is not possible to have less labour on the farm. Growth in business size which takes up this spare capacity can result in considerable economies.

Against these advantages can be ranged an array of possible disadvantages from large-scale production, again not all of which are relevant to farming.

Disadvantages of Large-Scale Production

(a) With increasing size a business often has to introduce some formal organisation. Rules and regulations become necessary. The close contact between manager and workers is lost and the cost of "unproductive" administrators may cancel out economies of scale.

---

* See also the advantages which it is suggested may accompany the establishment of monopolies in Chapter 4. These are often closely associated with those of large-scale production.

(b) The separation of management and workers lengthens the chain of consultation so that changes by the management cannot be implemented quickly and views, ideas and observations of the operators are not rapidly communicated to the managers. In contrast, on the vast majority of farms a worker can easily contact his employer if he notices something wrong with a crop or animal.

(c) Because large businesses are usually run by salaried managers and not the owner, the incentive for efficiency and hard work is diluted. This can be obviated by profit-sharing agreements or piece-work payment (e.g. some cowmen are paid according to how much milk a herd produces).

(d) Large complex firms demand a high standard of management ability. Lack of such talent is usually quoted as the most important limiting factor to the size of businesses. Overstretching mediocre talent can result in costly mistakes. An agricultural example might arise if a cowman could easily manage a 60-cow herd but made expensive oversights when the herd was raised to 100 cows.

(e) The division of labour (specialisation) often associated with large-scale production, for example in the motor industry, is alleged to reduce workers to the level of mere machine-minders with no interest in their product's quality and encourages industrial unrest. Specialisation also puts great potential to disrupt into the hands of a few key workers.

In addition to the above, the following factors contribute to the persistence of small-scale production:

(i) craftsmen-entrepreneurs take pride in the quality of their work and may not wish to expand; the utility which a violin maker using traditional methods derives from his activities comes in part from a satisfaction in creating beauty with his own hands, and he may eschew larger-scale factory methods even though they are technically feasible (and are used by some firms) for violin manufacture;

(ii) where personal service is essential e.g. water diviners or where hand craftsmanship is the prime characteristic of the product e.g. roof thatching;

(iii) where the market for the product is limited e.g. a dry-cleaning shop in a small village will tend to remain a small business; artificial legs are not demanded in vast quantities so large-scale production is not justified;

(iv) most large firms had their beginnings as small firms and some present small firms *may* be embryo large ones.

If expanding the scale of operations to gain certain advantages, say in marketing a product, brings with it diseconomies in the form of inefficiencies in the production process, firms can be expected to take steps to avoid these diseconomies. Hence, in the complex motor industry we find that some of the technical processes of production are sub-contracted out to other—and usually smaller—firms. Also, where diseconomies are suspected, manufacturers often set up duplicate production plants around the country so that the best scale of production is not exceeded, but the advantages of large-scale selling techniques can be used to dispose of all the identical products from the various factories. In agriculture, farmers form groups to gain the advantages of bulk buying or selling and co-operate in machinery-sharing syndicates; these forms of horizontal integration allow cost or price advantages to be enjoyed while still allowing the farms to remain largely independent businesses.

Is There an Optimum Scale of Operation?

The question is often asked, "Is there an optimum scale of production?" or, more specifically, "Is there an optimum size of farm?" To attempt to answer such a question, we would need to know the shape of the LRAC curve for the type of production under consideration. If this were U-shaped, with economies of scale merging directly into diseconomies and with no region where neither operated (as back in Fig. 5.30), under a situation of perfect competition firms would compete in price with each other until the product was produced at the lowest possible cost corresponding to the lowest point of the LRAC curve. All firms would be of the same scale—that scale whose lowest point on the SRAC curve corresponded to the lowest point of the LRAC curve. However, in the real world, competition is not perfect and technical innovation and changes in the relative prices of factors of production go on which constantly shift what could be thought of as the "ideal" size of the firms. In the industry which is probably close to perfect

Fig. 5.31 *The General Pattern of Economies of Scale in Farming*

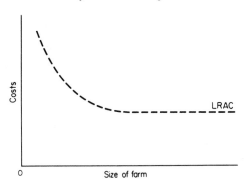

competition, agriculture, economies of scale are not vividly apparent beyond the size of farm requiring two or three men. For most types of farming in the United Kingdom (and probably also in the agricultures of many developed countries) the LRAC curve appears to be L-shaped as shown in Fig. 5.31. While very small farms can generally be shown to suffer from serious cost disadvantage on average, so that, when moving to larger sizes, considerable economies of scale are apparent, beyond the 2-3 man size* neither economies nor diseconomies seem to operate, even among the largest farms found in the UK at the present level of prices, costs and farming techniques. This could be one explanation why large, medium and relatively small farms can apparently compete successfully with each other and are all viable at the prevailing level of prices for agricultural products.

A feature of UK agriculture over the last few decades has been the general fall in the number of small farms and rise in the number of large ones, with a "break-even" point of about 300 acres. Farm enlargement in order to reap economies of scale no doubt forms part of the explanation for this movement. However, upward adjustments in the size of agricultural firms are often difficult to make because of the

---

* Discussions of economies of scale in farming frequently use farm area or estimated labour requirements as their measure of size rather than the volume of output; within the same type of farming they are reasonable proxies for output.

characteristics of the factors of production used by farming. Most notably, farmers often find it impossible to acquire additional land, and agriculture requires proportionally more land than most other industries. The nature of Factors of Production, with special attention to their use by agriculture, is discussed in the next chapter.

Exercise on Material in Chapter 5

Note: Students who have omitted the first part of Chapter 5 should commence these exercises at 5.4.

5.1 *The Production Function*

    (a) Draw the graph of a typical one product-one variable input production function, exhibiting both increasing and diminishing returns, with Average and Marginal Product curves. Graph paper with labelled axes is provided at the end of this exercise.

    (b) Using it, answer the following:

      (i) At the point of inflection of the production function .... Product is at its maximum.

      (ii) At the point where a line from the origin is tangential to the production function .... Product and .... Product are equal.

      (iii) When Total Product is at its maximum .... Product is zero.

      (iv) When total production is declining with increasing quantities of input, ... Product is negative and .... Product is positive and declining/increasing.

      (v) The point where Marginal Product is greatest is also called the Point of ....

      (vi) Where, apart from when only one unit of input is used, do Total Product and Average Product coincide? ....
....

5.2 You are given the following information about the production of a firm. Complete all the columns.

Output (units)	Total Revenue £	Marginal Revenue £	Total Costs £	Fixed Costs £	Variable Costs £	Average Marginal Costs £	Average Total Cost £	Variable Cost £
0			110					
1	50		140					
2	100		162					
3	150		175					
4	200		180					
5	250		185					
6	300		194					
7	350		219					
8	400		269					
9	450		349					

(a) Does the firm appear to be producing under perfect or imperfect competition?

....

(b) What are its fixed costs?

....

(c) Plot Marginal Revenue, Average Total Cost, Average Variable Cost on the same graph. Also plot Marginal Cost, but to get over the difficulty of "lumpiness" of the input, plot the Marginal Cost of each unit at its mid-point (e.g. plot the MC of the second unit of output half-way between one and two units on the horizontal axis).

(d) A firm is in equilibrium when *Average/Marginal/Total* cost equals *Average/Marginal/Total* Revenue. (Cross out the inappropriate words.)

(e) At what output is the firm at equilibrium?
(Take the nearest whole-number of units of output.)

....

(f) Estimate the profit (net revenue) earned by the firm at equilibrium in two ways:

(i) Total Revenue (   ) minus Total Costs (   ) =

(ii) Average Revenue (   ) minus Average Total Costs
(   ) multiplied by the number of units of output
(   ) =

(g) What is the minimum price of the product at which the firm will produce, and what will be the quantity produced:

    (i) in the long run?   Price ....

                                  Quantity ....

    (ii) in the short run?   Price ....

                                  Quantity ....

5.3 A chicken farmer can alter the composition of the food he uses for his flock by substituting barley for oats. If he increases both the yield of eggs increases. The isoquants he faces for two levels of production are given below:

230 eggs per bird/year		250 eggs per bird/year	
Tons of Oats	Tons of Barley	Tons of Oats	Tons of Barley
10	0.5	10	0.8
8	0.7	8	1.4
6	1.2	6	2.2
4	2.2	4	3.8
3	3.4	3	5.6
2	5.2	2	10.0
1	11.0		

(a) Plot these two isoquants.

(b) If the price of oats and of barley are the same at £30/ton, what is the least-cost combination of barley and oats the farmer should use in the food to produce 230 eggs per bird?

                         Tons of Oats ....

                         Tons of Barley ....

    (*Hint*: construct an iso-cost line and slide it until it is tangential to the 230 egg/bird isoquant.)

(c) What are the relative proportions of the two cereals in the food?

                         Oats/Barley = 1/ ....

(d) What is the combined cost of oats and barley in the food if they both cost £30/ton? . . . .

(e) What is the least-cost combination to produce 250 eggs/bird if the price of oats equals that of barley?

Oats . . . .

Barley . . . .

Are these in the same proportions as (c) above?   YES/NO

(f) What is the least-cost combination to produce 230 eggs/birds if the price of oats drops to half that of barley?

Oats . . . .

Barley . . . .

Does the fall in the price of oats result in more or less oats being used in the food?

. . . .

5.4 A farmer, with his resources on a one-hundred acre farm, can produce either all wheat or hay for sale, or various combinations of the two. The possible combinations are:

Tons hay	Tons wheat
0	200
50	250
100	260
150	200
200	140
250	50
260	0

(a) Plot his production possibility curve, with wheat on the vertical axis. Graph paper with labelled axes is provided.

(b) From the table estimate his marginal rate of substitution of products

(i) between producing no hay and producing 50 tons

. . . .

(ii) between producing 150 tons of hay and 200 tons of hay

. . . .

(c) What are the most profitable levels of wheat and hay to produce if

(i) the price of wheat and hay are the same per ton?

Wheat (tons) . . . .

Hay (tons) . . . .

(ii) that price of wheat is double that of hay?

Wheat (tons) . . . .

Hay (tons) . . . .

(iii) the price of wheat is treble that of hay?

Wheat (tons) . . . .

Hay (tons) . . . .

(*Hint*: construct iso-revenue curves to find these quantities by connecting quantities of hay and wheat which bring in the same revenue.)

(d) If the farmer knows that he will be unable to sell his hay at any price, will it still pay him to produce some? (Assume that it can be destroyed at no cost.)

YES/NO

STUDENTS WHO HAVE OMITTED THE FIRST PART OF CHAPTER 5 SHOULD START HERE

5.5 What of the following best describes a Fixed Cost in production economics terminology?

(a) The cost of any input whose price per unit has been fixed by long-term contract.

(b) The cost of any input which rises in a fixed proportion with output.

(c) The costs of production which would have to be borne even if the firm temporarily ceased production.

5.6 Cross out the inappropriate alternatives from the following statement:

"If Marginal Cost is rising with increasing output, Average Total Cost *will be falling/will be rising/may be rising or falling*. Average Fixed Cost will be *rising/falling*. The minimum price which the entrepreneur will be able to accept in the short-run will correspond to the lowest point on the *Average Variable Cost/Average Total Cost* curve, but in the long-run price must be at least the lowest point on the *Average Variable Cost/Average Total Cost* curve. The rising Marginal Cost curve cuts both the *Average Fixed Cost/Average Variable Cost* curve and the *Average Total Cost/Total Cost* curve at their lowest points. To make maximum profits an entrepreneur will select that level of output at which *Average Revenue/Price/Marginal Revenue* equates with *Average Total Cost/Average Variable Cost/Marginal Cost.*

5.7 (a) Give three sources of economies of scale

. . . .

. . . .

. . . .

(b) Give three sources of diseconomies of scale

. . . .

. . . .

. . . .

EXERCISE 5.2

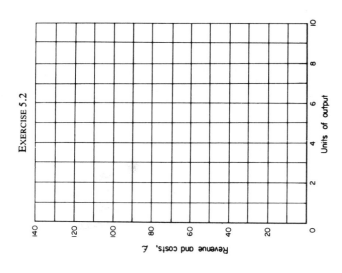

EXERCISE 5.1

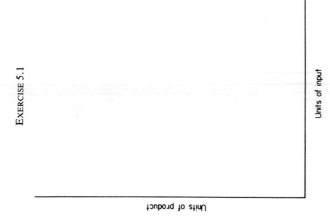

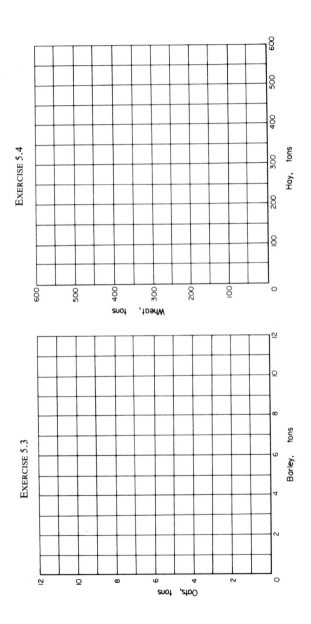

CHAPTER 6

Factors of Production and their Rewards (Theory of Distribution)

The Purpose of Production

Iᴛ is easy to lose sight of the reasons why production takes place when becoming engrossed in the technicalities of production economics, or the implications of elasticities of demand. This is an appropriate point to restate that economics is concerned with the study of how the wants which people have—for food, clothes, cars, hi-fi, entertainment and a whole host of other goods and services—are satisfied within the limits set by the resources which are available. This section examines these available resources in more detail.

The Nature of Production

The term "production" conjures up the image of factories making cars, freezers, clothing or similar goods and in the process employing workers, buying steel, paint, rubber, textiles and the host of other necessities. This picture is rather too narrow, but contains the essential nature of production—that consumers require the resources available for production to be changed into a form in which they can satisfy wants. In car manufacture, metal, glass, plastic and other materials are assembled by a labour force working in a factory into a product demanded by the consumer. This is one type of production process and each of the items needed in it can be described as a "Factor of Production". We will see later that factors can be grouped into four main categories.

Assembly is only one of the forms which production can take. Iron production from iron ore using fuel and blast furnaces changes the *form* of the metal; it is then passed perhaps to a second production process where its *shape* is changed by casting or pressing into components which may be the starting point for perhaps another process. Nor must production be limited to tangible goods: the provision of entertainment is just as much a production process. Although the music pounded out by a public bar pianist cannot be seen, it is the output from the combination of his physical effort, mental ability and his instrument (machine). Similarly, transport is itself a production process, as are also marketing and advertising.

Agriculture is fundamentally different from other industries because its production processes are dependent on biological growth, not manufacture.

Specialisation and Exchange

A man stranded on a desert island would need to satisfy his wants by producing his own food, shelter and so on; he would be forced by circumstances to become self-sufficient. In our modern society such independence is rare, although farmers are perhaps in an unusual position in that they can be more self-sufficient than most—they can feed their own family, produce their own timber for building repair or for fuel etc. Most people, however, participate in production by specialising in one task or occupation, such as being a secretary or teacher or factory worker for which they receive payment, and with this money buying all the other things which they require. A farmer does not build his own tractor as well as trying to raise animals and crops; instead he concentrates on his farming, sells his produce and with the proceeds buys the machinery he requires. At the same time, workers in the tractor factory do not grow their own food, attempting to be self-sufficient; they concentrate on tractor production, receive payment and spend this on food, clothing or whatever. As has been pointed out earlier, specialisation and exchange permit a higher level of production and consumption than could be achieved by self-sufficiency; the formal explanation for this will be presented when the benefits from international specialisation and trade are described (Chapter 8).

Classification of Factors of Production

All production processes, whether of the self-sufficient sort or the product-exchange sort, involve taking a collection of inputs (for milk production these would include animal feed, water, labour for milking and stock care, land to grow fodder, buildings and equipment, management skills), and combining them in some way, resulting in an output, or product. In our example we have several products—as well as milk, which is the primary product, calves and dung are produced as by-products. It is generally agreed that the inputs to production processes can be classified into four broad groups of Factors of Production—land, capital, labour and entrepreneurship.

Each main group of factors of production is considered below individually. Almost all production processes involve all four. Knocking a crab apple off a wild tree with a stick is a very simple process, but all four types of factor are present. The earth's resources (land) have produced the apple, the stick is a piece of capital, labour was involved in using the stick, and the decision to pick the apple was an entrepreneurial one.

Land as a Factor of Production

Land as a factor of production has a particular importance for agriculture because, compared with other industries, farming requires a lot of it. In the United Kingdom agriculture (excluding forestry) takes up about four-fifths of the total land area but only engages about three per cent of the population.* Just as we describe a firm which needs a high proportion of labour in its mixture of inputs (such as thatching, or any craft work where mechanisation is impossible) as labour-intensive, so we should label agriculture as land-intensive when comparing it to most other industries.

In economics the term "Land" not only includes the soil surface, but embraces minerals under the surface, water, climate, topography—indeed everything that is a "gift of nature" and not the result of man's

---

* The precise figure is open to dispute—should the percentage of the working population or total population be taken? Should workers in ancillary industries—like veterinary surgeons or animal feed firms—be included? The overall picture, however, is very much the same whichever figure is taken.

past activity. The early economists paid much attention to land as a factor of production, notably Thomas Malthus (1766-1834) and David Ricardo (1772-1823).* Land was considered special in that it was strictly fixed in quantity, and was thought the only factor subject to the Law of Diminishing Returns. We now know that neither characteristic is true—productive land can be reclaimed from the sea, or lost through coast erosion, and diminishing returns apply to labour and capital too.

Land still has, however, several distinguishing characteristics in modern economics justifying its separation from the other factors.

1. Land means *space* in which production processes take place. Almost all such processes demand some space—factory space for car manufacture, office space for a firm of accountants— and agriculture requires more space than most. Crops require space to grow, cows require space to graze, although the space requirement for egg production and pig rearing is much less than most other farming enterprises and this has earned them the title "factory farming". When the areas required to grow food for the housed animals are also taken into account, "factory farming" may not be such an appropriate label.

2. Land means *location*. Pieces of machinery can be moved from place to place, and even buildings can be taken down and rebuilt elsewhere (for example, some UK buildings have been transported and rebuilt in the USA), but a piece of land is actually the location it occupies.** Much of the value of a building site on farmland adjacent to an expanding town arises from the uniqueness of that land's location. J.H. von Thunen in *Der Isolerte Staat* (The Isolated State) of 1826 pointed out that, all other things being equal, farmland nearest markets would fetch highest prices because transport of produce to the market was shorter, and hence easier and cheaper. The further from the market one progressed, the lower would be the value of the land. Ease of transport and land values would mean that crops of high value per acre (e.g. horticultural produce) would be grown nearest the town, with less intensive production further

---

* Some of Malthus's theory is mentioned in the section on Economic Growth in Chapter 6, and Ricardo's approach to rent is touched on later in this chapter.
** Even carting away the soil from a patch of earth does not alter the location of that patch—a big hole will now occupy the spot.

out. This would produce concentric rings around the town of horticulture, then perhaps dairying, then perhaps corn and beef production (see Fig. 6.1).

Such a location theory breaks down in a world of easier transport (von Thunen wrote in the time of horse transport) so that the location of each crop tends now to be more heavily influenced by where the most suitable soils and climates are. Nevertheless, location is still influential in, for example, the growing of peas in areas accessible to freezing factories, or the organisation of large mixed farms where the use of a field for grazing rather than potatoes may be as much determined by the fact that it is within a cow's walking distance from the milking parlour as by its soil type.

3. Land is a *repository of natural forces*. This third characteristic of land is probably one which springs most readily to the minds of agriculturalists. Agriculture (using the term in its broadest sense) is unique in being so fundamentally dependent on natural forces—physical, chemical and biological—contained in the soil because it is concerned with biological growth rather than manufacture. Farmers are thus pre-occupied with "soil fertility" which to them is the practical manifestation of the sources of growth. Indeed, to them the term often acquires semi-religious overtones. The agricultural economist Edgar Thomas has clarified the term in the following way:

FIG. 6.1 *Land Use Based on von Thunen's Location Theory*

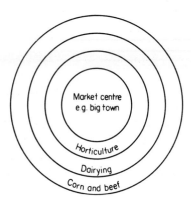

"Soil fertility includes all the climatic, ecological and other natural factors which determine how far land can be used for the production of crops or for the growth of livestock."

Soil fertility "is conserved under any system of farming provided such a system is technically capable of being continued indefinitely. Indeed, a system of production from land which does not conserve soil fertility should rightly be regarded as belonging not to agriculture but to mining".*

If land is "a repository of natural forces", then the term can easily be interpreted as including coal, oil and other minerals. Little land, however, remains untouched by human hand. Certainly in developed countries the countryside is a result of the "gifts of Nature", having been worked on by man to clear forests, drain marshes, plant hedges, improve soil structure and correct acidity and so forth. This past effort has resulted in improvement in levels of production, and forms a convenient link to the next factor of production to be considered—capital—because this past effort is really a form of the next factor.

Capital as a Factor of Production

The term "capital" means different things to different people. If a farmer were asked to list the items making up his capital, he would probably include the value of all the assets he owns and uses on the farm—his animals, machinery, buildings and land. He would probably deduct the size of any borrowings he may have made from banks or relations. A shareholder in a large public company might interpret his "capital" as the value of his shares; shares indicate a part-ownership of the assets of the business to which they relate.

To an economist the term "capital" implies something rather more precise. Goods like shirts, shoes, tea, bread etc. are wanted by consumers, and are hence called "consumer goods". Other goods, like tractors, machine tools, factory buildings and milking parlours are not wanted for their own sakes, but are wanted by producers because they help in the eventual production of consumer goods. Such articles are thus called "producer goods" or more commonly, capital goods, pieces of capital or, simply, capital.

A definition of "capital" might be "anything which has been produced

---

* Thomas, Edgar, *An Introduction to Agricultural Economics*, Thomas Nelson & Sons, 1949.

and is used to increase the effectiveness of current productive activity". Another is "wealth employed in the production of a flow of income". The essential nature of capital can be explained by a simple model. Imagine a factory producing cars (the nature of the product is not important), and imagine that this factory occupies in total 6,000 square feet of floor space (i.e. the factor "land" is present), employs 140 men ("labour" is present) and uses 100 machine tools of a variety of types ("capital" is present). Part of the factory (1,000 sq. ft. and 10 men) is set aside to make machine tools to replace the existing ones as they wear out, and the effort of these men is just sufficient to maintain the existing stock of machines. The factory produces 500 cars per year. The total output of the factory can be represented by the following equation:

Total production = Production of cars + production of machine tools
  (consumer goods)   (capital goods)

Suppose the management (i.e. the factor "entrepreneurship" is present as well) decides to increase the number of machine tools used in the factory, and that it does this by increasing the number of men and the floor area devoted to making machine tools. It will have to divert men and area from actually making cars to do this, but the rate of production of machine tools will rise so that they are being produced faster than they are wearing out. This means that the number of machine tools used on the shop floor will rise to a higher level, at which the increased production of tools just keeps pace with the wearing out of tools in use. With the greater number of tools, the output of cars eventually rises to 700 per year.

Before	*After*
Total labour force 140 men	140 men
Total factory space 6,000 sq. ft.	6,000 sq. ft.

Making cars	*Making cars*
130 men, 5,000 sq. ft. 100 machine tools	120 men, 4,000 sq. ft. 110 machine tools
Making machine tools	*Making machine tools*
10 men, 1,000 sq. ft. Output 500 cars/year	20 men, 2,000 sq. ft. Output 700 cars/year

Several points can be illustrated with this model:

(a) In the "after" situation, the same total labour force and floor space is used with a *greater quantity of capital*, with a resulting higher level of output. There is no *guarantee* that building up the stock of capital in use will increase output, but in this case the factory's investment in additional capital goods had had that effect, perhaps because the labour force can do things faster (say, with electric screwdrivers), or do things with machines which were previously not possible, such as lifting.

(b) In the factory as a whole there is a greater quantity of capital used relative to the quantities of the other factors, which in total have not altered; the production has become more *capital-intensive*. In the car-making part of the factory, fewer men are used but more machinery; not only has this process become more capital-intensive, but there has also been a substitution of capital for labour. (Note that the term capital-intensive has nothing to do with Capitalist as opposed to Communist. These are labels applied to types of economic systems to indicate who owns the stock of productive capital. If either type increased its stock of capital goods relative to the quantities of its other factors of production, then either could be described as having adopted more capital-intensive production methods.)

(c) A third feature of this factory is the *division of labour* (or labour specialisation). Some men specialise in producing machine tools while others specialise in car production. Although the model does not spell it out, the probability is that, within the two divisions of the factory, some men specialise in, say, electrical work while others use lathes. The division of labour has several obvious advantages. Workmen acquire skills if they specialise and the employment of people with specialised training or abilities becomes possible, e.g. designers or draughtsmen. In certain circumstances there may be substantial time savings, for example, if a blacksmith can keep his forge constantly in use, work will be done quicker than if individual workmen have to light up a forge each time they need one. Also there may be less fatigue if, for example, one man fits the bonnets to cars while another, working at a lower level, fits sumps; the alternative would be for one man to clamber up and down. Perhaps the most significant aspect of the division of labour is that breaking down the production process into small simple operations allows more mechanisation and maybe even automation, where the limitations of human abilities are replaced by untiring machines.

While the division of labour can be advantageous, it also has negative features. The monotony of repetitive acts can lower efficiency and encourage industrial unrest. Workmen can become alienated—they lose the craftsman's interest in his product and become mere machine minders, and a strike by a few key specialists can halt the output of complete factories.*

(d) the fourth feature of the factory model is that of the period of *waiting* which elapses between the time when the decision to build up the stock of machine tools is made and the time when the output of cars rises to its new higher level. This is illustrated thus:

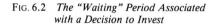
FIG. 6.2 *The "Waiting" Period Associated with a Decision to Invest*

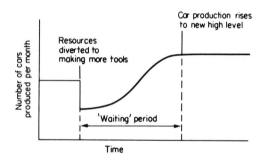

Note that when some men and floor space are switched from car production to tool production the output of cars falls and only recovers after the new tools have been constructed and are in use. With the enlarged number of tools the *eventual* level of car output is higher than at first, but the factory's operators have had to put up with a fall in output in the short-run to make possible a high final output level.

(e) The fifth feature of the model is that of *maintaining capital intact*. This has already been touched on, but is of sufficient importance to stand as a separate feature. Capital goods wear out with time and usage—this is called capital consumption. While this is obvious with machines and vehicles, it also applies to roads, buildings and so on. Capital goods are only wanted by producers because such goods aid

---

* See also the association of specialisation with the economies of scale cited in Chapter 5.

production. Clearly, a new machine with a full working life in front of it is capable of aiding production for a lot longer than a worn-out machine, so the value of the new machine to the producer is much greater than the old one. In our model, if steps were not taken to compensate for the wearing out of machine tools, after a time the factory would have found itself equipped with worn-out and hence valueless capital. Some of the factory's resources were, however, devoted to producing replacement tools, so that the overall value of machines in use did not fall (or, in other words, the value of the total capital in use was maintained intact). Note that this meant in the short run the output of cars suffered because men who could have produced cars were making tools, although this higher level of output could not have been maintained indefinitely because the machines would have been getting older. An alternative policy for the factory (although this implies a different economic model) might have been for it to use all its men and floor space for making cars, and to set aside out of its income a sum to enable it to buy new machine tools from other factories specialising in tool-making.* This sum, called a depreciation allowance, could be calculated by knowing how fast machines wear out, and hence when replacements would be needed.

Capital in the Real World

Before passing to the next factor of production, it is worth broadening the discussion of capital to the model's implication for the real world. The less-developed countries are keen to expand their levels of production and thereby raise the living standards of their populations. To do this they require more capital (factories, roads, etc.) yet they are caught in a vicious circle of poverty which prevents them building up their stocks of capital. Investment requires some resources to be diverted to the production of capital goods, yet if all a country's resources are already fully stretched in providing necessities for life, then little or none can be spared for capital build-up unless extreme sacrifices in living standards are made. In contrast a developed country would find it much easier because the general level of production is higher to start with. This is illustrated in the next page.

---

* This implies a greater total quantity of factors of production used in car manufacture by the extent of those used by the tool-making firms.

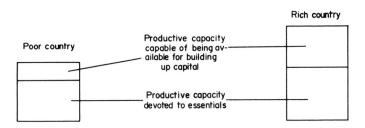

Even though in the rich country more production per head may be considered necessary to satisfy "essentials", this still leaves a greater margin which could be used to build up capital (or to produce "luxuries").

The vicious circle of poverty can, however, be broken—at least in principle—by a loan from the rich to the poor for the purchase of capital equipment. This enables output to grow, resulting in a more-than-proportional increase in the poor country's margin for investment. Technical advice, which permits a greater output from given resources, also increases the margin. So would, in theory, aid in the form of consumer goods (e.g. milk powder) which should allow a country's own productive capacity to be devoted to capital production. The danger in this last method is that it tends to depress the local consumer goods industries without a corresponding increase in capital production. The poverty cycle is illustrated, together with ways of boosting the circle, in Fig. 6.3a.

A parallel situation exists in the small farm in the UK and other developed nations. Low outputs from small farms means that there is little left for investment once the living expenses of the farmer have been deducted from the farm's income. Low investment in livestock, machinery, buildings and so on will keep the farm's output low, and so the cycle continues. In the UK attempts have been made to break this cycle by the injection of capital through grants to cover a proportion of the cost of new buildings and equipment, by giving free technical advice through the Agricultural Development and Advisory Service, and by maintaining incomes through guaranteed prices for farm products or subsidising costs of production (see Fig. 6.3b).

It would be wrong to leave our discussion of capital without a reference to the banking system and the "capital market", the latter

FIG. 6.3a   *The Vicious Circle of Poverty*

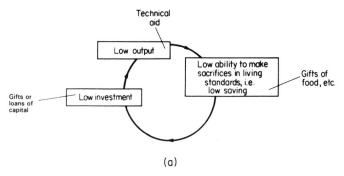

(a)

FIG. 6.3b   *The Small Farm*

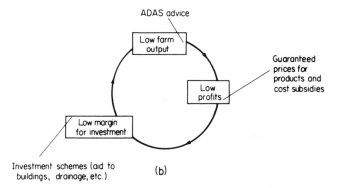

(b)

being the market for new or existing stocks and shares. In our factory model we envisaged capital as being built up within the production unit. Some countries have built up their capital in this way—notably the USSR which enforced great hardship on its population during the 1920s and 1930s in terms of low living standards so that industries could be built up. At the farm level, many UK farmers pride themselves on having built up their businesses by self-denial and without recourse to borrowing.* It must be recalled that much of this went on during and

---

* It appeared that in 1969 just over half the farms in England had no liabilities other than the short-term deferments of payment until the end of accounting periods, a practice widespread in commerce. More recent evidence suggests that, despite a greater level of indebtedness in the late 1980s, most of the total borrowing is accounted for by a small minority of farms.

immediately after the Second World War when prices of farm produce were kept high to encourage production. However, there is often no need to depend exclusively on internally-financed capital. The banking system enables entrepreneurs who have opportunities to use extra capital profitably to acquire the necessary machines etc. by using purchasing power granted by the banks, who in turn base their lending on deposits made with them by people who have the ability to buy goods, but do not wish to do so. Public companies can raise funds by issuing *new* shares which members of the public or investing institutions can buy in the "capital market", for example through the London Stock Exchange.

Labour as a Factor of Production

A parallel can be drawn between capital and labour in that a country's productive ability will be related to the quantity and quality of both of these factors. Labour in production is possibly best thought of as Labour Service, since it is what men and women do which is their contribution to production and for which they receive payment. The *quantity* of labour service available in a country will depend on the size of the population, and the proportion of it which is available for work at any time. This in turn will be affected by the country's attitude to such things as whether women should be expected to work, the ages of school-leaving and of retirement, the age and sex structure of the population and the commonly accepted length of working week and holidays.* The factors influencing the supply of labour service are altered as the country's standard of living rises and the general trend is to reduce the proportion of the population in the work-force with greater affluence (school-leaving age rises, retirement age falls; the proportion of the population above retirement age rises—even the general attitude to work probably changes). However, there are also movements in the other direction; many more UK wives and mothers now go out to work than was customary in the 1950s. It might be noted that in the UK the length of the working week for agricultural workers is on average some 1½ hours longer than for workers in other industries

---

* Further interesting parallels with capital might be drawn; like machines, workers' age, but do they wear out? A build-up of capital is similar to a rise in the working population.

and their earnings are lower (about three-quarters of the average earnings in other industries).

The quality of labour service determines how productive the labour force is. A highly efficient labour force usually means a high level of output per worker. This is affected by the physical well-being of the work-force, and a certain minimal standard of health, clothing and food is required below which efficiency will fall dramatically. Working conditions are influential, as can be witnessed by the output achieved when riddling potatoes in a rain-and-windswept clamp as opposed to the output in a protected barn. A well-trained and educated work-force is conducive to high output; hence we have seen the setting up of various industrial training boards in the UK. The degree of capital intensity is important—one man can be more productive (in terms of output per man) on a powerful tractor than on a less powerful one, as is also the quality of the other factors with which labour is combined. A trained worker hindered by poor equipment on poor land and managed by an inept farmer cannot hope to produce a great deal.

Many of the factors which can affect efficiency are often considered two-edged. Education* to read and write, or a technical training, can improve output per man, but the results of a more advanced education can be counter-productive if it encourages doubt about whether increasing production should be the overriding goal of society. Whether more advanced education is considered to be productive or counter-productive depends on what the objectives of education are. Similarly, social services covering unemployment and sickness benefits can promote efficiency by providing a minimum standard of health and living during misfortune, but some people would argue that social services, by removing anxiety associated with dismissal and unemployment, also remove the spur to efficiency. However, the removal of anxiety may in itself be a goal of society, and any deleterious effect on the output of goods and services may be considered a small price to pay for this increase in well-being.

An interesting hypothesis on the relationship between well-being and productivity has been put forward by the Chinese. Communist China

---

* Part of education can be considered an investment just as purchasing a machine-tool might be an investment. To educate people, resources have to be used, and after a period of "waiting" these resources may well reap a reward in terms of the increased productivity of the educated people.

was criticised for the wasteful use of its manpower in the 1950s to 1970s, because people there were expected to switch between jobs—manual and mental, urban and rural, in leadership and on the shop floor. The Chinese justified this partly on the grounds that individuals cannot be considered fully developed unless they have the opportunity to express all facets of their ability through their work, which implies that workers should have experience of a range of jobs, and partly by the claim that a population of well-developed, fulfilled and happy members will in the long run be *more* productive in terms of output per man. This view retreated in the 1980s. Specialisation and financial incentives have become major ways of achieving greater production, especially in agriculture where private farming in pursuit of profit is now encouraged.

Entrepreneurship as a Factor of Production

Land, labour and capital do not produce anything unless they are organised together. Even knocking crab-apples from a wild tree requires the bringing together of the tree (land), and stick (capital), and an arm to wield the stick (labour). Furthermore, the decision to knock the crab-apple must be made, and the risk taken that the attempt may be unsuccessful and hence the effort wasted. The person who takes the decision to gather the crab-apple, organises the factors of production and bears the risk is the entrepreneur. In a farm firm he will be responsible for deciding what is produced (barley or oats etc.), how production is implemented (e.g. whether to use labour-intensive or capital-intensive methods), the choice of the scale of production (say, two 50-cow dairy herds or one 100-cow herd) and for marketing the product (when, where and how to sell). All this requires an ability to make and implement decisions and the larger the business generally the higher the qualities that are demanded. People differ in their organisational ability and their ability to assess risks, so the quality of entrepreneurship is highly variable. While some economists would maintain that the organisational role of the entrepreneur is just a rather specialist form of labour, the risk-taking element is a unique characteristic of the entrepreneur in a capitalist economy.

When discussing entrepreneurship, we most often take as a typical example the operator of a smallish firm who is both its owner and its manager. If he launches an enterprise after assessing the risks involved and it proves successful, he benefits through reaping profits. If his endeavours are ill-founded and flounder he will make a loss. The vast majority of farms in the UK (and Western Europe) are arranged like this, with a sole proprietor of perhaps two partners bearing the organisational and risk-taking functions, so the convention is acceptable for a study of economics associated with agriculture. However, it must be borne in mind that agriculture is exceptional among industries in capitalist economies. Most other industries consist of large firms where the people responsible for managing the firm and taking the decisions (the directors) are not necessarily the people who suffer if the decisions prove to be wrong. The chief losers will be the shareholders of the firm. In the case of nationalised industries the individual or group who make decisions is probably difficult to identify, and the ultimate source of important decisions may be the elected Government. When decisions prove wrong, then presumably everyone suffers to varying extents. In being mainly concerned in this text with agriculture in Western capitalist countries with market economies we can count ourselves fortunate that in these the farming entrepreneur is relatively easily identifiable, although in a study of agricultural decision-making in a socialist country with a centrally planned economy the identification of the body taking the entrepreneurial decisions might be much more difficult.

The Origins of Risk and Uncertainty

Risk and uncertainty arise because production is not an instantaneous process. Time elapses between when an entrepreneur decides to go into a line of production and when the product finally reaches the market. The entrepreneur has to assess costs and prices and physical performances before he embarks on his enterprise, and these may change with time. Taking the case of barley production, once a farmer has planted his seed he is fairly heavily committed to the crop for the year. The cost of his inputs (e.g. fertiliser, insecticides, fuel etc.) may rise unexpectedly, the price he finally gets for his barley may be affected by things totally beyond his control (such as a change in the attitude of Russia towards buying grain from the USA, upsetting the expected prices on the international barley market), and his physical production may be affected by drought or excessive rain or uncontrollable disease. Agriculture is unusual (although not unique) in having its levels of output heavily influenced by climatic conditions. Not so unusual, but of major importance, is the guiding hand of Government policy which can often change more rapidly than the industry can adjust. For example, a Government policy may be set to encourage beef production; the time required to produce a beef animal from conception to slaughter weight will be generally not much less than two years, so a lag can be expected between farmers deciding to respond to the incentive for more beef and the increased supply reaching the market (called a Supply Lag). During this time, however, Government policy may have been changed and beef prices may have fallen well below the high prices originally envisaged. Some supply lags are even longer—in blackcurrants for example a period of about three years elapses between planting and the harvest of the first crop. As has been pointed out in Chapter 4, the combination of long supply lags and a lot of relatively small producers acting independently are conditions conducive to cycles of low and high prices.

In addition to the factors already covered, entrepreneurs have to contend with risks that their capital equipment may be destroyed by fire, flood etc. before its useful life is completed, that it may become obsolete by advances in technology (such as happened to the stationary baler when tractor-drawn machines were developed), and that they may themselves suffer illness or accident, jeopardizing the viability of the enterprise.

Difference Between Risk and Uncertainty

A conventional distinction, although by no means rigorously adhered to, is that unwanted happenings can be divided into two categories; those for which insurance with an insurance company can be arranged are termed risks (e.g. fire, personal accident), while those for which no formal insurance is available are termed uncertainties. Formal insurance is based on a statistical analysis by insurance companies of, say, the known number of farm fires each year, and premiums worked out on the basis of the total cost of claims, with a suitable margin to give the company a profit. The farmer benefits because, for a small, known, certain cost (the premium) a large, unknown, possible loss can be avoided.

The classification into risks and uncertainties is by no means absolute. While farmers widely use some fire or personal insurance (i.e. fires and accidents are usually regarded as risks) it *is* possible to insure against rain (as is done by some village fetes) and some diseases, although the premiums are generally beyond what many farmers considered justified— they prefer to treat such occurrences as uncertainties and take alternative avoiding measures.

Avoidance of Risk and Uncertainty

The entrepreneur can take steps to minimise the effect of risks and uncertainties on his business. Risks can be insured against, while a wide variety of steps can be taken against uncertainties. The *choice of enterprises* is important, as some are prone to uncertainty. Beef production is so prone because of the long production cycle and the possibility that the returns, coming in in a single lump sum when the animal is sold, may coincide with a time when prices are depressed. Others, such as milk production in the UK, are traditionally safer because a cheque from milk sales comes in at regular monthly intervals from the Milk Marketing Board and prices are known within a relatively small margin of error well in advance.

Another method of combating uncertainty is *diversification*, i.e. the combination of several enterprises on the same farm e.g. cereals, beef and grass-seed production. This reduces the overall uncertainty because the likelihood of all the enterprises suffering from, say, price falls at the same time is smaller. On the other hand too much diversification introduces the possibility that neither enterprise will be well managed, and perhaps most of the benefit from diversification on UK farms is available by the time the number of major enterprises has risen to three. Enterprises which contrast in some way are good—perhaps being harvested at different periods of the year to minimise the effect of adverse weather. Such enterprises may have the additional benefit of spreading out the farm's labour requirement, avoiding seasonal peaks.*

---

* In UK agriculture, many farms are operated by entrepreneurs who have business interests in other industries; farming is part of their diversification. About one third of the total number of farmers has another source of earned income, predominantly from other businesses run in parallel with farming. Part-time farmers can be found on all sizes of farm. Part of the Common Agricultural Policy's approach to the problems faced by European farmers, resulting from falling product prices, is to encourage them to start enterprises which would not normally be considered as part of agriculture, for example by providing tourist accommodation, setting up craft industries in their buildings, or embarking on forestry on their farms.

A third method is *production flexibility*, which can take several forms:

(a) *Cost* flexibility implies adopting production methods which have high variable costs and relatively low fixed costs. A suitable example is calf-rearing, which can occur either in purpose-built, environmentally controlled rearing houses, with efficient use of food (i.e. high fixed costs in the form of interest on capital in the building and depreciation but low variable costs), or in cheap straw-and-corrugated-iron structures with relatively inefficient use of food (low fixed costs, high variable costs). Total costs in each case might be the same but, in the event of the price of reared calves falling, the operator with the cheap structure could easily terminate the enterprise by salvaging the corrugated-iron for some other use and setting fire to the straw, whereas the operator with the expensive building would be forced to continue production or even expand to recoup as much of his original outlay as possible from his reduced profit per animal (see also the discussion on the "reverse" supply curve in Chapter 3 and the decision to continue production in the short run and long run in Chapter 5). Cost flexibility is also given when substitution between variable inputs is possible; a glasshouse heating system that can readily be switched from oil fuel to gas and back again will be less subject to uncertainty caused by price changes and supply interruptions than a system solely dependant on one fuel.

(b) *Product* flexibility is achieved when a firm's resources can readily switch products according to changes in product prices. A good example is the Friesian dairy herd which can either produce dairy replacements or beef animals, depending on the type of bull used, according to whether the price of young dairy stock or beef stores is the more attractive. A herd of Hereford beef-type cows, in contrast, cannot produce animals suitable for dairying, only for beef production. Another example of product flexibility is afforded by the wide span, general purpose, floored and walled farm building which can be used for on-floor grain storage, housing dairy or beef animals, machinery storage or (if the farm is turned over to leisure use) as a covered tennis court. Such a building would fare much better in times of uncertainty than, say, a specialised poultry house which can only be used for one product.

(c) *Time* flexibility is achieved by choosing those enterprises which have a short production cycle. All other things being equal, enterprises with long lags between initiation and fruition (e.g. the establishment of a plum orchard) will be more uncertain than those with short lags (e.g. cereals, or even radishes) because the longer time period gives more opportunity for changes to occur. Another form of time flexibility is exhibited by products which store easily; a farmer is given a wider choice of where he markets his output than with a highly perishable product.

A fourth general method of countering uncertainty is by the use of *informal insurance*. This may take the form of keeping an extra tractor on the farm beyond what would be considered normally necessary to act as a safeguard against unexpected breakages at critical times, or having a larger than normally justified combine harvester as an informal insurance against bad weather necessitating rapid harvesting. Pesticides are sometimes applied without the offending insects being apparent, just in case they are there and develop later. Farmers often err on the generous side when applying fertiliser because, while too little may reduce yields noticeably, a little too much will not do harm. A common practice by farmers is to carry over a bay of hay from one year to the next in case the crop in the following year is unexpectedly light, and a type of informal insurance which is probably more legendary than fact is the mattress stuffed with money which farmers are reputed to keep as a precaution against crop failure or disease. The more modern manifestation of this legend is for farmers to maintain their bank accounts unnecessarily in credit or to arrange an overdraft limit and then stay well within it.

A fifth method of avoiding the deleterious effects of uncertainty is by arranging *contracts* for inputs or ouputs. If a farmer can arrange with his supplier of animal feedstuff a contract to buy a certain quantity over a year at a given price he will be insulated from subsequent rises in feedstuff price, at least to the end of the year. Similarly, a calf producer may contract to sell all his calves to a rearer at a fixed price for a certain time period, and is thereby protected against falls in the market price of calves. If the cost of inputs and/or the price of products are known in advance, planning the business to operate at its most profitable level becomes much easier. Many of the precautions listed above which can be taken to minimise uncertainty become no longer necessary (such as diversification, informal insurance, cost flexibility), and the firm's resources can be utilised more efficiently.*

Vertical integration is said to occur when two or more stages in a production chain fuse together. Contracting is a form of vertical integration, but the term is more often used when a change of ownership is involved. An example would occur if a farmer or group of farmers bought up a retail butchery business and sold their own meat through it. While the motives for such a take-over are complex, one of them could well be the desire to have a secure outlet for their product, and to be no longer reliant on prices offered by other butchers. Another example would be where a supermarket bought a horticultural holding so that it could be sure of a reliable supply of produce of controlled quality for its shelves, thereby reducing its risks. Yet another example of vertical integration is the Co-operative Society movement which operates farms with dairy herds, large collecting dairies, a distribution system and retail milk rounds.**

In addition to the ways of reducing uncertainty already discussed, entrepreneurs have an increasing amount of *market intelligence* upon which to base their decisions. While in other industries this may partly come from privately-commissioned research, in agriculture it comes very largely from the Ministry of Agriculture which publishes a wide range of data which is available to farmers. (If they used it more perhaps price cycles would be greatly reduced!)§

---

* Economic theory suggests that the reduction in risk must be paid for in some way. For example, the contract price which a blackcurrant grower might negotiate for a ten-year period with a manufacturer of blackcurrant cordial could be expected to be a little below the average of prices on the open market for the same period.

** Note that *Horizontal Integration*, where firms at the same stage of production band together, is mainly to form powerful bargaining groups. For example, farmers forming a buying group can buy in bulk and obtain considerable discounts.

§ For a few commodities which are produced on UK farms (notably wheat and barley) a "futures" market exists, i.e. where buyers and sellers take options to purchase at some date in the future, although they may in fact possess no grain to sell or may have no intention of buying. However, such a "futures" market enables precautionary transactions to be made to minimise the effect of real price changes. This process is called "hedging" and is not widely used, if at all, by individual farmers in the UK, though it is a feature of the activities of traders in agricultural commodities and animal feedstuffs manufacturers.

Mobility of the Factors of Production and Unemployment

Factors of production can be considered mobile or immobile in two senses—the occupational sense and the geographical sense. This classification according to mobility runs across that of factors into land, labour, capital and entrepreneurship, so we can find occupationally immobile capital, geographically mobile labour and so on. As with most systems of classification, some factors are found which do not happily fit into any single category, or which fit into several according to circumstance, but this method of describing factors is useful because it helps to explain why payments to the owners of factors vary. This will be developed later.

Occupational Mobility

Occupational mobility refers to the ability with which a factor can be switched to alternative uses. Factors which are occupationally immobile i.e. where they cannot be easily switched to an alternative use, are often called "specific", while occupationally mobile factors are also termed "non-specific". A milking parlour is a specific (or occupationally immobile) type of capital because it cannot be used for anything except milking cows, and if the farmer decides to produce milk no longer, the parlour will simply go out of use. In contrast, a covered yard, intended for dairy cattle, can be used easily for a wide range of other farming activities—beef housing, machinery or hay storage, even perhaps as an indoor tennis court if desired; such a building would be considered non-specific, or occupationally mobile.

Similarly some types of land are restricted in the crops which they can grow; heavy clay soils in areas of high rainfall are probably limited to permanent grass production. Such land is "specific". On the other hand, a medium loam in a more kindly climate can be switched between a variety of crops, so is less specific.

Some farmers feel particularly at home with, say, sheep-rearing and perhaps have little knowledge or ability with other types of farming. They would be considered specific (or occupationally immobile) entrepreneurs, whereas a farmer willing and able to vary his enterprises or even embark on a totally different type of business would be relatively non-specific.

The classification of labour into specific (occupationally immobile) and non-specific (occupationally mobile) is not quite so straightforward. Generally, specific labour implies that for which a long period of training or a special ability is required. People with special skills or training are generally unwilling to move to other occupations—a veterinary surgeon

FIG. 6.4   *Examples of Factors of Production Classified According to Factor Group and Occupational Mobility*

	Occupationally immobile	Occupationally mobile
Main group of factor	(Specific)—factors cannot easily be switched to alternative production processes	(Non-specific)—factors easily switched to alternative production processes
Land	Heavy clay in wet area—suitable only for permanent pasture Very sandy soil in dry area—suitable only for crops capable of withstanding dry conditions Land covered with a motorway—cannot easily be changed to farming or other uses	Medium soil, well drained, in area of equable climate—wide variety of crops Park in central London—can be used for buildings or recreation or farming
Capital	Milking parlour—only suitable for original purpose Blast furnace, submarine	General purpose farm building—farming uses or non-farming use Estate car—can be used in a wide variety of firms for carrying goods, or for pleasure
Entrepreneurship	Elderly sheep farmer—only capable of managing one type of farm	Young businessman willng and able to embark on a wide range of business activities
Labour	Medical labour, musicians or other artists—unable or unwilling to transfer to other occupations Older skilled or semi-skilled labour unable or unwilling to retrain for another job	Workers able and willing to switch jobs Workers able and willing to switch jobs

or a specialist dairyman or a concert pianist is usually unwilling to take a job which does not use his skills, even though he could easily do the non-skilled job. At the other extreme, a general labourer could find a job in the building industry, on farms or even working in shops—his labour is non-specific and he can switch between labouring jobs easily. We must not try to put a veneer of perfection on our classification. Specialised labour can usually do non-skilled jobs, but its skills are then wasted and its earnings often lower. Sometimes this switch out of a skilled occupation is made necessary by technical advance—the number of skilled hand-spinners of wool needed in the textiles industry was greatly reduced by the introduction of machinery during the British Industrial Revolution. Similarly the number of farriers required on farms was reduced by the introduction of the tractor to UK agriculture. On the other hand, it is very difficult to switch labour *into* a skilled job, either from the pool of non-specific labour or from other types of specific labour when demand for the skill increases. This has implications for the earnings of people in the sought-after job, and these will be discussed later.

Fig. 6.4 is a table combining the two types of classification which have been discussed so far.

Geographical Mobility

Geographical mobility implies the ability of the factor to move from place to place. With the factor "land", the question of geographical mobility hardly arises, except that if "land" is used to include all natural resources, then items like coal and water can be made to move from place to place.

Some units of *capital* (e.g. tractors) are geographically mobile, but many (most buildings) are not. However, it must be recalled that even immobile capital wears out, and the money allowances which are made to compensate for this need not be re-invested on the same site—replacement structures can be erected elsewhere—so that in the long term capital *is* geographically mobile.

The geographical mobility of *labour*, which is the willingness and ability of workers to move from place to place, is significant not only from the restricted view of shifting this productive resource (labour) from where it is plentiful to where it is in short supply, but because geographical mobility has major social implications. As a broad generalisation, workers who have undergone long periods of training in non-manual skills, and are therefore termed "specific" in the occupational sense, are willing to move from area to area for jobs; teachers are generally willing to move for promotion, and the "brain drain" of medical staff and scientists from the UK to North America is also illustrative of geographical mobility. Also as a broad generalisation, non-specific labour is generally less willing to change its location. Factory workers in declining industry in the north of England, particularly the more elderly, may not be willing or able to move to the south where jobs are available. To assume that this immobility of the "blue-collar" worker is the result of an over-attachment to a particular area is a gross over-simplification of the case; the desire to stay in familiar surroundings, close to family and friends may be an entirely rational pattern of behaviour and in line with economic principles once it is recognised that satisfaction comes from both monetary and non-monetary sources. The *ability* to be mobile is highly relevant; high income earners are more able to finance moves, especially if this involves buying a house. Lower earners, even if they are houseowners, may be constrained if considering a shift to an area of relatively high house prices. To ease mobility employers may have to offer relocation incentives, such as a subsidy on mortgage payments for the first few years following a move or a generous lump-sum which enables new carpets and furnishings to be afforded. Labour dependent on rented accommodation will be unwilling to move unless suitable housing to rent is available at the new location. It seems that such accommodation is often scarce in places where job vacancies exist for blue-collar workers. For national policy two alternatives then present themselves; firstly, construct new housing to let where jobs exist and, secondly, encourage new industries to establish themselves in areas where labour (and housing) is available. Both have been tried in the UK, as was illustrated by the Coal Board building houses to encourage miners to move to expanding collieries, and by the designation of parts of the country as "development areas" including some Rural Development Areas to which industries are encouraged to go by financial incentives.

In agriculture the mobility of hired labour (both occupational and geographical) is closely linked to housing, the worker's age and the availability of alternative employment. Often labour is deeply committed to an area and is unwilling to move to a different farm or type of farming. However, specialists such as dairymen or managers are probably more geographically mobile. When young workers wish to marry, geographical mobility may rise because they will tend to take a job where a house is available with the job. On the other hand, once established in a tied cottage, the ability to leave farm work for more attractive employment in another industry may be seriously hampered if such a move necessitates leaving the house. In a period of high house prices, alternative accommodation may be impossible to rent or buy, i.e. occupational mobility is constrained. No doubt the arguments for and against tied housing, with its political connotations, will continue for a long time.

Unemployment of the Factors of Production

Unemployment of the factors of production is closely related to their mobility, although immobility is not the sole reason for unemployment. While labour unemployment is the type which most readily springs to mind, unemployment of the other factors occurs and can be described in a similar manner. There are various main types of unemployment and they differ in their cause.

Mass (or cyclical) unemployment is the form associated with the slumps which tend to alternate with booms in most of the Western economies. The 1929-35 Great Depression was a particularly severe example. From 1975 the UK has experienced yet another period of unemployment of this type, related to the efforts to curb inflation. The cause of mass unemployment is a deficiency in the general level of demand in the economy, and this is correctable by Government action, although other goals of the Government may have to be sacrificed in its pursuit. This type of unemployment will be met again in the section on macroeconomics.

Structural unemployment is the type to which factor mobility bears a direct relevance. It occurs when the demand for labour or the other factors engaged in particular uses such as in the mining industry diminishes. If the factors are occupationally or geographically immobile they will be unable to find alternative uses, so will remain unemployed. One cause of structural unemployment can be a reduction in the *demand* for the product of firms, as occurred when miners were made unemployed in the 1960s because the demand for coal fell (and hence the demand for miners). At the same time labour may be in short supply in other industries because of an increased demand for their products. In other words, this unemployment arises from the continually changing structure of demand in the country. Another cause of structural labour-unemployment could be the development of new *labour-saving techniques* which require fewer workers. New machinery means that UK farms can be run with a much smaller labour force than twenty years ago, and this has caused some farmers to shed labour. In areas with few other industries, rural unemployment can be a serious problem.

Historically workers have often been afraid that new machines would threaten their jobs, and the Luddites (active from 1811-1818) set about destroying spinning machines to try to stop mechanisation of the textile industry in the UK's Industrial Revolution. However, in practice new jobs have in the past been created which have absorbed the unemployment either in the machine-making industries or by the development of other industries which have prospered as the result of the economic growth resulting from mechanisation. If factors are willing to move from place to place and to change their occupation (i.e. if they are both geographically and occupationally mobile), technical unemployment will be greatly reduced.* While this has been true of the UK, it does not necessarily apply to the present-day less developed countries, where the mechanisation of agriculture and other industries can, if taken too fast, worsen unemployment if alternative employment opportunities are not created at a sufficient rate. Even with developed economies, the fear among some is that labour-saving technologies, devised and adopted at least in part as a response to the increasing bargaining strength of unionised labour, will create a growing pool of unemployment which will have a serious impact on the stability of society and which can only be obviated by changes to the economy, including the reform of labour unions.

Some occupations are essentially of a seasonal nature, and this can give rise to *seasonal unemployment*. Apple-picking and strawberry-picking take place over relatively short periods, and some gypsies or "travellers" take advantage of their geographical and (within a range)

---

* See also the section on economic growth in Chapter 8.

occupational mobility to move from farm to farm and crop to crop as their services are required, thus obviating seasonal unemployment and removing the need for farms to carry large "regular" staffs to cope with seasonal peaks in the demand for labour which are characteristic of many types of agriculture.

To complete the picture, there will always remain some unemployed people, even when the demand for labour is high. This *residual* unemployment will consist of those who are incapable of being employed for physical or mental reasons or those who do not wish to take a job. People on strike could also be included as unemployed; their unemployment is "voluntary" but may cause others to be laid off and produce some involuntary unemployment. In addition, the process by which workers become informed of job vacancies, apply and are selected, takes time. Better communications could reduce this "friction" in the labour allocative process as could, some suggest, lowering unemployment benefit to increase the necessity of taking a job. The term *frictional unemployment* is frequently used to describe unemployment resulting from the failure of labour to flow smoothly between uses.

From the individual factors themselves we must next turn to how they are allocated among the range of production processes in order to generate the goods and services which consumers demand.

The Allocation of the Factors of Production and their Rewards

The definition of economics that has been cited in our introduction to the subject (Chapter 1) was that "Economics is the study of how men and society choose to allocate scarce resources between alternative uses in the pursuit of given objectives". If the objective is known, then there will be some allocation of the resources (or factors of production) which enables that objective to be best fulfilled. For example, assuming that the objective is to maximise the value of the nation's production (begging the question of what this implies precisely), then labour, land, capital and entrepreneurial ability could be allocated between industries and firms in such a way as to achieve this. In the section on production economics (Chapter 5) we saw that this optimum allocation of scarce factors between a range of alternative uses will be achieved when each of the factors is so allocated that the addition to production caused by the last unit (the marginal unit) of each factor in each use is of the same value. This is an example of the use of the Principle of Equimarginal Returns. Take for example the allocation of vehicle drivers between industries. Being a driver in one industry requires very much the same sort of skills and abilities as in any other, and switching between industries is easy. If the last man in manufacturing industry results in a greater value of increased output than the last man in agriculture, then there will be a net gain if drivers are transferred from agriculture to manufacturing. Diminishing returns will apply so that a point will eventually be reached where the value of the marginal product of labour in each industry will be the same and no net gain would come from further switching of drivers. Any allocative process should be aiming to achieve this optimum pattern of resource use.

In a centrally-planned socialist economy the deployment of the nation's factors of production between different industries and locations is decided and directed by some form of central authority. It will direct labour from one industry to another, and capital from place to place. It will be attempting, perhaps not always consciously, to equate marginal returns to factors in each use. Provided that the central planning authority knows what types and quantities of final products its population prefers, what resources are available to it for production, and the technical nature of the production functions in each line of production (to give it an indication of marginal products), such an allocative procedure can result in the optimum pattern of resource use—at least in theory. In practice, the lack of knowledge of consumer preferences, available resources and technical coefficients hampers the system and considerable quantities of resources may be used up simply in providing the planners with the information they require to form decisions. In a basically capitalist market economy the allocation of resources is achieved largely through the price mechanism. We have seen in Chapter 3 that the wants of consumers are reflected in what they are willing to buy, i.e. the demand they express in the market for goods and services. If, for example, consumers develop a liking for colour television sets, this will be reflected in a strong demand and waiting lists for sets. Some consumers will be willing to pay more than the existing prices. In order to meet this strong demand, the producers of sets will need to expand their output; to do this they must attract labour, capital etc. from producers of other goods. This they do by offering higher rewards than other producers, creating a reward differential which they can afford because the strong demand for their products enables them to raise the price of sets somewhat and still sell all the sets they can make. The prices of consumer goods and prices of factors of production are thus linked, and signals of consumer demand (the prices of consumer goods) are passed on by producers to the market for factors of production in the form of factor prices, so that factors move between lines of production, attracted by differentials, to try to best satisfy consumer demand. Recall, however, from Chapter 4 that the price mechanism contains imperfections which may necessitate some modification to the allocation of factors it produces, e.g. the growing of opium may need banning.

Demand and Supply of Factors of Production

In a free market economy both a supply of factors (land, labour, capital, entrepreneurship) and a demand for these factors exist. The demand for factors is said to be a *derived* demand because it comes from businesses which produce goods that consumers demand. For example, the demand for dairy cows is derived from the consumer demand for milk. When the demand for factors and the supply of factors are allowed to interact, a price results. The price of labour is the wage rate, the price of land service is rent and similarly, the price of entrepreneurship is profit. Shifts in the demand and/or supply curves for factors will affect prices in a similar way as experienced with consumer goods, and a similar set of imperfections arises when there is a sole buyer of a factor (i.e. a monopsonist), a sole supplier (as typified by a trade union operating a closed shop), when one firm or the whole industry is considered, and so on. The market for factors is by no means completely smoothly operating, and the failure by Government to recognise its shortcomings, particularly the resistance of labour to falling wages, was a major contributor to the high unemployment during the 1920s and 1930s in the United Kingdom. This will be returned to in Chapter 8. However, some aspects of the demand and supply of factors must be considered here.*

---

* The derivation of the demand and supply curves for a factor, and how changes in product price and productivity relate to the quantities demanded and supplied, require a somewhat more advanced treatment than is used in most of this text. The reader is referred to one of the standard textbooks, listed on p. 337, for these areas of economic theory.

The responsiveness of the demand for a factor to changes in its price is, of course, the *price elasticity of demand* for that factor. This elasticity will depend on a number of variables:

(i) *The price elasticity of demand of the product.* If the price of a factor increases (e.g. animal feedstuffs) the price of the product (e.g. milk) will tend to rise. If this results in a severe cut-back in the demand for milk (i.e. milk has an elastic demand), then the demand for animal feedstuffs will be curtailed heavily too.

(ii) *The availability of substitutes.* If good substitutes exist, as happens when closely similar animal feedstuffs are available from rival firms, then a rise in price of one brand will cause much less of it to be used as farmers switch to the alternatives i.e. demand is elastic. Contrast this with what would happen if the prices of all brands were increased simultaneously.

(iii) *The relative cost* of the factor to the total costs of production. If the cost of water represents only a tiny proportion of the cost of milk production, then an increase in the cost of water will not cause farmers to use much less of it. In contrast, feedstuffs form a major proportion of the total costs of egg production and any cost increases will affect total costs to a significant extent.

The responsiveness of the supply of factors to changes in factor prices is dependent both on the time period in question and on whether the supply to individual firms or the national supply is under consideration. Nationally, the stock of land is almost completely fixed, irrespective of its sale price or rent, but the amount of it being supplied to the market will be sensitive to price, and an individual farm firm can increase its stock of land by outbidding its competitors. With capital a similar situation obtains: nationally the stock of capital in the form of plant, machinery and buildings is fixed in the short run at least, but single firms are not necessarily so restricted and can acquire more, changing the distribution of the national stock. The national supply of labour *can* be increased by paying higher overtime rates to encourage longer hours, or by tempting women from full-time housework to join the labour force in industry, or by delaying the retirement age. Individual firms have, in addition, the opportunity to tempt workers from competing firms by offering higher rewards.*

---

* Note the reverse supply curve, cited in the section on Supply in Chapter 3, which can occur when the length of working week has risen to a level at which further wage rises cause *fewer* hours to be worked as people prefer to take more leisure and the same money income.

Dynamic and Equilibrium Differentials

The movement of factors of production from firm to firm or from industry to industry has been explained in terms of their owners switching them in order to benefit from higher rewards, thereby tending to iron out differentials. This type of differential is a *dynamic* one because it is temporary, and is eroded by movements of labour, capital and so on. Not all differentials, however, are of this type. For example, the pay of university staff is below their equivalents in industry; part of the explanation is that university teaching has some non-monetary compensations, such as greater time flexibility and freedom of expression. This differential is an *equilibrium* differential; removing it by raising university salaries to industrial levels would cause a flood of applications from outsiders for university posts. Land in city business centres can command a much higher annual rent than agricultural land in the countryside because of the higher earning power of its location and its scarcity value. Such a differential will be permanent because it is impossible to increase the quantity of land in a city centre.

Economic Rent in the Reward to Factors of Production

"Economic rent" is the term used to describe the differential element in the reward which the owner of an economically scarce factor of production receives above the reward which could be earned in the factor's best-paid alternative use. Thus the term "rent" is used here in a way rather different from its commercial meaning. While economic rent was first described by reference to land, it can be applied to each of the groups of factors, although the term "quasi-rent" is then sometimes used.

Economic rent can apply to any factor which is less than completely elastic in supply, at least in the short run. Economic rent can be defined as
THE MARGIN (OR SURPLUS) WHICH A FACTOR EARNS ABOVE THAT PAYMENT NECESSARY TO ATTRACT IT INTO OR KEEP IT IN ITS PRESENT EMPLOYMENT.

A factor of production will be kept "in its present employment" as long as it cannot earn more elsewhere; its earnings in its best-paid alternative use is its *transfer earnings,* so we can redefine economic rent as "the margin (or differential) a factor receives above its transfer earnings".

To illustrate the concept of economic rent we will first take an example using the factor labour. If the horticultural industry faces a supply of labour which is less than infinitely elastic, as illustrated in Fig. 6.5, and it wishes to expand the number of persons working in horticulture from q_1 to q_2, then it will need to pay higher wages to attract workers from other forms of employment. Assuming that the same wage has to be paid to all workers in horticulture, then a man who was just willing to stay in horticulture at the old wage level, because his earnings just equalled what he could have earned elsewhere, now finds his rewards greater than his transfer earnings. At wage level w_2 labour will flow into horticulture from other industries, but this flow will cease at the point at which the last man who switches loses as much in terms of foregone earnings as he gains from horticultural wages. For this marginal man, his horticultural earnings are of the same size as his earnings in his best-paid alternative employment (i.e. his transfer earnings) but for all the other horticultural workers whose earnings exceed their transfer earnings, part of their reward consists of economic rent (i.e. a surplus above the payment necessary to keep them in their horticultural employment). The size of the total economic rent accruing to labour at wage level w_2 is shown by the shaded area in Fig. 6.5 and the amount of transfer earning in its reward by the unshaded area below the supply curve. From this graph it can be deduced that the shallower the supply curve (i.e. the more elastic or price sensitive is the supply of the factor) the smaller will be the proportion of economic rent in its total reward. Conversely, the more inelastic the supply, the greater the proportion which is economic rent, and if the supply were completely inelastic with the supply curve a vertical line, all the reward would take the form of economic rent.

The elasticity of supply of a factor of production can vary with the time scale under consideration and this can have consequences for the economic rent part of a factor's earnings. For example, during the 1960s UK dentists as a group seem to have received higher earnings than their abilities would have probably earned them in other professions—in other words, their actual earnings were above their transfer earnings, so that a sizeable part of their remuneration consisted of economic rent. This state of affairs had arisen because of a sharp rise in demand for dental services without a corresponding increase in the number of dentists. The high earnings attracted potential entrants to the profession

FIG. 6.5 *Economic Rent and Transfer Earnings in the Reward to Labour in Horticulture (hypothetical)*

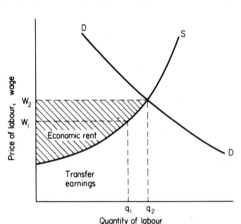

but considerable time was required before they could be trained, necessitating the expansion of dental teaching hospitals. Under such circumstances the highly inelastic short-run supply of dental services resulted in the profession earning considerable economic rent. The Government could have taxed dentists heavily in the knowledge that, as long as they were still earning more than in other professions that were open to them, they would stay in dentistry. In the longer term, economic rent could be expected to be reduced as the new entrants were trained and became qualified. This, combined with a reduced demand for treatment stemming from better awareness of the importance of dental hygiene among children, seems to have eroded dentists' high earnings. However, they have been replaced as a group enjoying economic rents by accountants whose large earnings arise from a strong rise in demand for their services and an inelastic short-term supply.

In the case of the accountants, the economic rent part of earnings may be eroded by the eventual arrival of "new" accountants to increase the supply, but with some factors of production no increase in supply is possible. If a pop singer earns £250,000 a year because of his unique abilities (i.e. there is only one person with his qualities), and his best-paid alternative job as a building labourer would earn him only £10,000 a year, the economic rent element of his present employment is £(250,000 − 10,000) = £240,000. He will not give up being a pop singer unless his earnings fall to the level of his alternative job.* The Government could tax away almost all the £240,000 "surplus" in his earnings and he would still remain a singer. This economic rent element of his earnings is what he receives for his uniqueness; any increase in demand for his services would raise his income and the economic rent content of his earnings, while a fall in demand would have the reverse effect.

Turning to the factor land for an example, we can cite the case of the high rents paid for land in city centres. As land for office development, a city centre plot might command, say, an annual rent of £10,000 per acre. If the next-best paid use of the land were for farming, let at £200 per acre, (i.e. its transfer earnings), then its economic rent would be £(10,000 − 200) = £9,800. This is an extra reward the land earns simply by being where it is—in the city centre. The number of sites in an existing city centre is absolutely fixed, so the overwhelming proportion of the land's reward will always consist of economic rent which can never be eroded by creating more sites; this is another case of an equilibrium differential which could be taxed (and has been in the form of rates paid by businesses) without changing the use to which the land is put.

Normal Profits, Surplus Profits and Monopoly Profits

When economic rent and the entrepreneur is discussed, the term "normal profit" is encountered. This term has already been introduced, notably in the chapter on competition (Chapter 4). Normal profit is the level of profit which an entrepreneur requires to compensate him for the uncertainties of production. If he is not rewarded to the level of normal profits, he will transfer his ability to some other line of production. Because the level of uncertainty differs between types of production, some being more hazardous than others, it is not surprising that the level of profit which is considered "normal" also varies.

If the entrepreneurs in, for example, beef production are just earning

---

* Non-monetary rewards of both jobs are ignored here.

"normal" profits and then the demand for beef goes up because of, say, an increase in the incomes of consumers, beef producers will be earning more than "normal" profits, and this surplus above the "normal" level is a form of economic rent. It is the surplus above the payment necessary to keep the entrepreneur in beef production. This surplus will attract entrepreneurs from other industries, more beef will be produced, the price will fall somewhat and the entrepreneurs will each be earning only normal profits again.

The profits made by a monopolist through his manipulation of the market have something in common with the economic rent accruing to the owner of a factor of production, since the monopolist also receives a reward greater than would be necessary to keep him in his particular line of production. The distinguishing difference is, however, that while economic rent is thrown up by the market system for factors, a system which is outside the control of individual factor owners, the monopolist deliberately contrives his surplus profit by exercising his power over the market through controlling the supply of goods.

The Functions of Profit

Profit has been mentioned a large number of times in the text so far in addition to its discussion immediately above. To conclude this chapter on the factors of production we will summarise the functions of profit in a free enterprise, perfectly competitive capitalist model of an economy. *Firstly* profit induces entrepreneurs into production by acting as a compensation for the uncertainties involved. *Secondly,* levels of profit above or below the "normal" level indicate to entrepreneurs which industries or product lines should expand and which should contract to reflect the changing wants of consumers. *Thirdly,* profits in excess of the "normal" level enable entrepreneurs to obtain the resources they require for expanding production by outbidding competitors from other industries. Production is thus expanded in these lines for which consumer demand has risen. *Fourthly,* profit encourages the most efficient firms in each industry (hence benefiting the consumer) by enabling the most profitable firms to expand by outbidding the less profitable firms. In that the more profitable firms are those which have lower costs of production, this means that the consumer will be able to purchase his requirements at the lowest possible prices. In such a model of the economy profit can be seen as a useful mechanism in ensuring that productive resources are allocated in the way which serves consumers' interests best.

Within agriculture in Western market or mixed economies, a good case could be made that profit performs these four functions adequately. The numbers of inefficient and unprofitable small farms have decreased rapidly since the Second World War. A striking example of the concentration of production has occurred with poultry meat, where many small units have been displaced by fewer very large producers and at the same time the real cost of this meat to consumers has fallen. However, it must be recalled that farming is unusual among industries in developed economies in that it is still largely comprised of many relatively small independent units, each of which has little power over the market. Where some degree of monopoly power exists, or where the State plays an important part in fixing prices, the existence of profit becomes far less obviously beneficial and the role it plays in the mechanism of allocating scarce factors of production in the best interests of consumers may be considerably reduced. Nevertheless, during the 1980s, Governments in the UK and USA have seen the pursuit of profits by competitive private firms as a vital ingredient in promoting an efficient and growing economy, and have made changes to encourage this, such as the privatisation of some previously-nationalised industries in the UK, such as telephone, gas, and British Airways. Concurrently, personal profit sharing and similar financial incentives have been introduced in the USSR and China to stimulate productivity.

Appendix to Chapter 6

A Trade Union in a Perfectly Competitive Industry

If workers in a perfectly competitive industry, to which farming approximates, form a Trade Union and successfully impose a minimum wage on the industry, the employers will be faced with a new supply curve for labour (see Fig. 6.6). The new supply curve has a kink at the level of minimum wage since, however few men the industry would wish to engage, they cannot be paid less than the minimum. If the minimum wage is fixed at or below the wage level which would result from the unhindered demand for labour interacting with supply, no

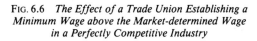

FIG. 6.6   *The Effect of a Trade Union Establishing a Minimum Wage above the Market-determined Wage in a Perfectly Competitive Industry*

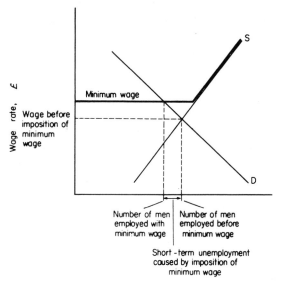

effect on the numbers of workers employed will result. One benefit which might occur in practice could be that unscrupulous employers would not be able to exploit the less well-informed workers, and the average wage might therefore rise a little.

However, if the Union attempted to raise the minimum wage above the equilibrium level, some short-term unemployment would result (see Fig. 6.6). These unemployed workers might well be willing to work for less than the minimum the Union was trying to impose, in which case the strength of the Union would be undermined, the non-wage benefits which could stem from unionisation be lost and low-wage exploitation reappear. In such circumstances legal enforcement of minimum wages might prove necessary and justified. In the United Kingdom the Agricultural Wages Board establishes legal minimum wage rates for farm workers after hearing representations from the

Trade Unions (principally the agricultural section of the Transport and General Workers' Union) and the farmers' Unions. Only about one third of farm workers are members of unions and this points to the difficulty of unionisation where the labour force is geographically dispersed and employed in units of small numbers of workers. Added to this the frequently close relationships between employees and employers and the involvement of livestock work against union militancy. Tied housing has also been influential.

The activities of Trade Unions in industries which are dominated by a few large employers require a different economic model for their explanation, a model outside the scope of this text. Reference should be made to one of the texts listed on p. 337. However, it is increasingly felt that the success of Unions in many industries in raising wages has been at the longer-term expense of increasing the level of unemployment in the economy, as employers have sought ways of substituting capital in production processes for the increasingly expensive labour.

Exercise on Material in Chapter 6

6.1 Why is the demand for a factor of production a derived demand?

. . . .

6.2 Give five separate goods from which timber derives its demand.

. . . .

6.3 (a) If the demand for glass-houses is elastic, will this tend to make the demand for glass elastic or inelastic?

. . . .

(b) The price of water makes little difference to the demand for it for bread-making. Why?

. . . .

(c) The demand for individual brands of fertiliser is more/less elastic than the demand for fertiliser in general. Why?

. . . .

6.4 Delete as appropriate; A farmer given a certain quantity of farm labour, must so allocate it among his various enterprises that the *first/last* man-hour used in each enterprise yields *the same/different* values of output. This will of course *mean/not mean* that the same amount of labour must be devoted to each enterprise.

6.5 Farmer Giles has only 8 days left for cultivating his land before he must plant his crop of potatoes. According to the best estimates he can make, his yield of potatoes from each of his three fields to be sown will vary depending on how much cultivation time is devoted to it according to the following table.

Days of cultivation	Field A Tons of potatoes	Field B Tons of potatoes	Field C Tons of potatoes
0	63	40	65
1	78	60	75
2	88	70	82
3	96	79	88
4	100	85	89

Use the Principle of Equimarginal Returns to devise the best use of his 8 days. At this optimum, how many days will he spend on:

Field A? . . . .

Field B? . . . .

Field C? . . . .

6.6 Farmer A is being displaced by a reservoir. He is willing to take a farm anywhere else in the country but is reluctant to change from sheep farming.

(i) Is he (a) occupationally immobile . . . .

(b) geographically immobile . . . .

(ii) Is his labour *specific/non-specific?*

(iii) Which is the more geographically immobile, specific or non-specific labour?

. . . .

(iv) Which is normally the more geographically immobile, specific or non-specific labour?

. . . .

(v) Which of the following are the more specific factors?

(i) *Motor van/road roller?*

(ii) Covered yard/cow kennels?

(iii) Moorland/medium/loam?

(iv) Old farmer/young man just leaving school?

6.7 Economic rent: a good actor earns £100,000/yr. In public relations he could earn £20,000/yr.

(i) Define economic rent ....

(ii) As an actor (a) what part of the payment he receives is transfer earnings?

....

(b) what part is economic rent?

....

(iii) Give two reasons why the actor receives economic rent for his services. ....

....

....

(iv) In the long run training schools for actors can be established which will increase the supply of actors. Why is it likely that good actors will always be paid above their transfer earnings?

....

....

....

6.8 Entrepreneurs' dual functions are the organisation of production and the bearing of risk and uncertainty.

(i) How may the entrepreneur remove risk from production?

....

(ii) Give two sources of risk in agriculture ....

....

(iii) Give three sources of uncertainty in ....

agriculture   ....

....

(iv) Give four ways in which entrepreneurs can reduce the effects of uncertainty.

....
....
....
....

6.9 Cross out the inappropriate alternatives.

(i) When prices and costs are stable an industry in perfect competition will so arrange itself that its entrepreneurs are earning *good/bad/normal* profits. If the beef-producing industry is in such an equilibrium, and the demand for beef suddenly increases because of an increase in consumers' incomes, in the short run beef producers will earn *surplus/normal* profits. New entrepreneurs will be attracted into the industry and force *up/down surplus/normal* profits until a level is restored where entrepreneurs are again earning *good/bad/normal* profits.

(ii) Define "normal profit". ....
....
....

(iii) Is interest on capital included in normal profit?   *YES/NO*

(iv) What happens if an entrepreneur earns less than normal profits in one industry?

....

(v) An entrepreneur would require a higher normal profit in the pop record business than he would require in the car retailing business. What is the chief reason causing this phenomenon?

Chief cause: difference in ....

CHAPTER 7

Some Problems of Using the Market as a Resource Allocator

Aт several points in the text so far it has been pointed out that the pattern of national consumption, production and resource allocation which would result from the free interaction of supply and demand for goods and services may need modifying if the best interests of society are to be served. Examples of this happening in practice are not difficult to come by. The consensus view is that addictive drugs such as opium should not be available on the open market, as would almost certainly happen if the State did not ban its possession and sale. On the other hand, the State provides some goods and services, such as education, without direct charge to the recipient, and uses legislation to compel children to "consume" education until they reach minimum school leaving age. The cost to UK farmers of some forms of farm capital, such as slurry handling equipment, is lowered by grants, thereby encouraging a greater level of investments, while in contrast the price of consumer "luxury" items is frequently raised by a higher rate of Value Added Tax than is applied to items such as food (which in the UK is zero rated). Some utilities such as rural buses and country pharmacies receive subsidies out of taxation and can thereby charge a lower price than they otherwise might. Clearly the view leading to these modifications to prices and quantities must be that the pattern of consumption and production which an unhindered price mechanism would result in is not the optimum and that a preferable pattern, preferable that is from the view of society as a whole, can be obtained by adjusting the solution provided through the market to the fundamental problem of satisfying the wants of consumers from the resources which are available. Although the price system can be useful to an economy in indicating the pattern of consumer demand, where current production leaves shortages and surpluses, and in channelling resources into satisfying demand, it also contains imperfections which should be recognised and understood. This interlude attempts to provide some explanations of why society acts to overcome the imperfections and paves the way for a study of the working of the whole economy in the next chapter.

Adam Smith and the "Invisible Hand"

The father of modern economics, Adam Smith, wrote in his famous book *Wealth of Nations* (1776)* the following: 'Every individual endeavours to employ his capital so that its produce may be of greatest value. He generally neither intends to promote the public interest, nor knows how much he is promoting it. He intends only his own security, only his own gain. And he is in this led by an invisible hand to promote an end which was no part of his intention. By pursuing his own interests he frequently promotes that of society more effectually than when he really intends to promote it'.

The doctrine implied in this quotation of leaving the economy to its own devices, known as *laissez-faire*, we now realise can only result in the "best" way of using the nation's resources under a very special set of conditions which do not exist in reality, certainly not in the sort of economy found in Western developed countries. One condition is that perfect competition should exist throughout the economy and another is that the actions of producers or consumers should not be felt by others in the form of "externalities"; this latter point will be developed later. Because these conditions are not fulfilled in practice and for a variety of additional reasons, it is not desirable to pursue a completely *laissez-faire* policy if the object is to promote the general well-being or happiness of society, commonly described as society's "welfare" or "social welfare". While there are considerable differences of opinion as to how far the State should interfere with the market mechanism, most economists would agree that there are valid arguments for *some* modification to the price system's solution to the allocation problem on the grounds of improving the welfare of society.

---

* The full title of Smith's work is *An Inquiry into the Nature and Causes of the Wealth of Nations.*

The Value Judgements of Society

Even if the market system were totally devoid of imperfections and the "Invisible Hand" of competition worked in the way Adam Smith envisaged it, there is no guarantee that society would find the resulting pattern of production and consumption to its liking. On the contrary, many people would want to change things on the grounds of improving society's welfare and would attempt to do this through the political system. The sort of goals they might envisage as being desirable characteristics of society would probably include the provision of reasonable medical care for everyone irrespective of how able they are to pay, the right for children to be educated according to their several abilities and irrespective of the size of their parents' income, the relief of unnecessary anxiety from being out of work and everyone's right to have a roof to live under. The unhindered market system would not necessarily bring these desired goals to fruition. It might bring about "economic efficiency", but this will not automatically result in a "fair" or "equitable" pattern of consumption.

The price mechanism responds to people's purchasing power, their command over the market, so that goods will be produced only for those members of society with money to spend; the greater a consumer's purchasing power, the greater will be his influence in the market for goods and services and consequently the greater will be the proportion of national resources used in satisfying his wants. The price system makes no moral judgements on the desirability of the existing pattern of purchasing power, which is a reflection of the present distribution of income and wealth in the community. However, members of society do make moral judgements on the desirability of situations such as may be instanced by the ability of a rich man to buy medical care for his pet cat while the children of a poor man may be denied attention through being unable to pay; in order to provide medication for the poor children it might be necessary to tax the rich man and society might feel this to be justified on the grounds of improving the happiness of society as a whole.

In effect a redistribution of real income is being brought about and the proportions of total national productive resources accounted for by the rich man and the poor man are being made a little less unequal. Attempts at quantifying improvements in social welfare resulting from such actions, when some people are made better off while others are made worse off, are beset by theoretical objections and complications. However, the general political attitude, in which society's value judgements are presumably reflected, even if imperfectly, suggests that among other things equality of opportunity and a basic minimal standard of living are two goals to be striven for in the process of increasing the welfare of society. Hence in the UK we have striven for a reasonable degree of equality of opportunity for students to attend college or university irrespective of the size of parental incomes and have largely achieved it through the system of grants. Pensions for retired people are paid to assist their incomes once their ability to earn has receded so that a minimum living standard is guaranteed, together with the preservation of the freedom and personal dignity which comes from financial independence. The system of grants and pensions, financed by a taxation structure which falls more heavily on those with higher incomes, is in reality a mechanism for transferring command over goods and services from the rich with their more ample spending power to the poor, i.e. a redistribution of the income pattern thrown up by the price system. Because wealth is linked with income, the holders of large concentrations of wealth are liable to be subject to attempts by society to tax it away; in addition a more even spread of wealth is seen by some to be itself a desirable goal because of the influence and social status which wealth gives. In UK agriculture the rising value of farm land, determined by supply and demand where the influences are by no means restricted to those coming from the farming industry, has lifted the owners of farms into one of the wealthiest sectors of the community and has therefore brought them increasingly under the influence of society's policy of striving for a more equitable wealth distribution through capital taxation, a policy which some would claim is contradictory to simultaneous attempts to promote an efficient and productive farm sector.

Imperfect Competition

If we put on one side questions of social value judgements and externalities, a situation of perfect competition among producers and among consumers can be shown to maximise utility. The case for perfect competition can be summarised as follows: if perfect competition were to exist throughout an economy the prices that would result would correspond to the marginal cost of producing goods. They would also correspond to the utility or benefit which consumers attached to the marginal unit of each commodity. Because marginal costs indicate the quantity of productive factors used to produce the marginal unit, factors which could have been used to produce marginally more of other types of goods, we can say that under perfect competition the quantities produced will be such that the utility derived from the last unit of each good is just equal to the utility which might have been given by the goods which have to be foregone in its production. If we wanted to produce more apples than would result under perfect competition the benefit in terms of extra utility derived from the additional apples would be less than the utility lost through producing fewer pears or beef or whatever has to be foregone. In this simplified model total utility will thus be maximised by the perfect competition solution to the problem of how much of each commodity should be produced.

We have already seen that this "optimum" pattern which would be produced by perfect competition may need changing to correspond with social value judgements, and in practice perfect competition throughout the economy is not a reality. Monopoly or oligopoly is common, and under these imperfect conditions prices do not necessarily correspond to marginal costs of production. Monopoly is quite likely to develop where an industry's cost structure is dominated by fixed costs, for example railways, gas and electricity where fixed equipment such as a gas distribution system is very expensive, yet the cost of producing each extra unit of product (cubic metre of gas or rail journey) relatively small. Under such a situation, price competition tends to result in a small number of rivals, each operating at a loss with eventually only one survivor who then is in a position to exercise monopoly power. Alternatively, the rivals may come to an agreement between themselves and collectively act against the consumer by raising prices. If the reduction in competition results in a rise in price and a distortion in the pattern of production most desired by consumers, then society can be shown to be worse off. In Fig. 7.1 some curves are drawn which represent the combinations of two goods, food and beer, which are equally preferable to society—these are social indifference curves (I_1, I_2 and I_3). Also illustrated is the production possibility curve showing the various combinations of food and beer which can be produced under perfect competition (PP). Society is clearly best served—or to use

Fɪɢ. 7.1   *The Production Possibility Boundary
and Social Indifference Curves*

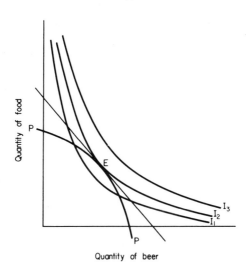

the conventional term, its welfare is maximised—at point E. This is the
one combination of products within the range of possibilities which
just touches curve I_2: no higher indifference curve can be attained. At
this point the number of units of food society is willing to exchange for
a unit of beer (which is the inverse of the relative prices of the two
goods) is the same as the number of units of food sacrificed when the
production of beer is increased by one unit. If a monopolist takes over
the beer-making industry, the anticipated result will be that he restricts
the output of beer and raises prices in order to maximise his profits.
This will cause the production of goods to move away from E, at best
by shifting around the production possibility boundary or retracting to
some point within it, so that society is moved to a lower indifference
curve and hence a lower level of welfare.

We have pointed out in Chapter 4 that a reduction in competition,
even to the extent of the emergence of a complete monopoly, is not
something to be decried automatically, since positive features may
emerge including economies of scale not considered in the simple

model used above. However, society keeps a careful watch on situations where the diminishing degree of competition could act against the public interest and applies a range of measures to influence the market outcome. These measures include legislation which restricts large-scale mergers between firms (the Monopolies and Mergers Commission encountered in Chapter 4), price controls over essential goods and services (though during the 1980s former regulations applying to many, such as milk, bread and public transport have been removed) and extend to full nationalisation. In its most complete form, nationalisation implies that the State not only owns all the productive units but also operates them, usually through agencies. In Britain the National Health Service is not only paid for out of public funds, but hospitals are also run as public organisations. While the historical reasons behind the nationalisation of industries are complex, especially if those firms which are completely or partially in public ownership but which are run as private firms are included, e.g. British Petroleum, part of the reason for keeping the "public utilities" such as gas, electricity and water under this form of operation from the 1940s to the 1980s was that this prevented the development of private monopoly while at the same time permitting the enjoyment of economies of scale. The statutes setting up nationalised industries commonly contained instructions that they were to be run "in the national interest". This has proved difficult to interpret in terms of the prices which should be charged and the extent to which unprofitable but socially desirable services should be provided at a loss. The 1980s saw a return to private operation of a string of former nationalised industries—telephone, gas, electricity, British Airways, British Road Services, even water supply—in pursuit of better efficiency. The danger of exploitation of consumers is countered by encouraging competition (for example, between gas and electricity as forms of energy, and by allowing new firms to compete with British Telecom in the telecommunications market) and by setting guidelines, with the possibility of Government intervention if breached.

Externalities Associated with Production and Consumption

A further reason why the price mechanism is imperfect as an allocator of productive resources from society's viewpoint is that it fails to take into account *externalities*. Externalities occur when actions of an individual through his consumption of goods and services, or a firm through its production processes, impinge on other firms or people in a way which is not reflected in market prices, so that the consumer or firm does not take into account in his decision-making these external effects he is having on others. Externalities may be beneficial, as when a farmer installs beehives in his orchard at his own private cost to pollinate his apple trees and where the bees stray and perform the same function for adjacent orchards and gardens. The neighbouring commercial orchards are thus receiving a benefit in terms of increased yields without cost to them and society in general has its enjoyment enhanced by more productive gardens. These bees are described as bestowing an *external economy* on others, but this will not influence the decision of the farmer to install the hive. His action will be based on the cost of the bees and the resulting improvement in yield from *his* trees. Another example of an external economy might be where a firm operates a training scheme for its employees and where these are free to leave and join rival firms which do not run such schemes. Without bearing any training costs these rival firms reap benefits of ready trained recruits. Similarly, when a government improves a coastal road for defence purposes local businesses and hotels may benefit because of better access for trade and tourists. Yet another example is provided by the farmer who invests in a better drainage system for his land to improve its ability to be worked and to enhance yields; the land of adjacent farmers may also benefit and their profits increase. In all these cases the benefit to society in general is greater than the benefit to the individual who undertakes the initial spending. This difference between private and social benefits has implications for the optimum allocation of resources.

Take, for example, Farmer Bloggs who spends money on improving his field drainage. A range of expenditures would be possible depending on the thoroughness of the drainage system and the better the system the higher the crop yield which might be expected over a run of years. We can assume that diminishing returns apply so that the additions to yield resulting from each extra £1,000 of drainage equipment installed decline as the total amount of expenditure increases. In deciding how simple or elaborate his drainage system should be, Farmer Bloggs will balance in his mind the extra cost of the more complex systems with the extra benefits they will give in terms of higher crop yields and revenues from his land over the expected life of the system; for him the optimal, most profitable system is where the marginal cost is just balanced by the marginal revenue (benefit). Both the costs and benefit are "private" to him since the effects on other farmers have not been taken into account; their extra yields do not benefit Farmer Bloggs and are irrelevant to his investment decision. However, if extra yield enjoyed by the surrounding farmers *is* taken into account, the wider benefit to society (marginal social benefit) of the field drainage can be assessed, and this is greater than the private benefit to Farmer Bloggs.

Just as there are external economies which cause a divergence between private and social benefits, so there are external diseconomies which cause private and social costs to diverge. External diseconomies arise when the actions of firms or individuals impinge deleteriously on others and are not taken into account by the decision-maker. A favourite example is the brickworks' chimney which belches dirty smoke, imposing "external" costs on others in society by causing housewives to wash clothes and curtains more frequently, offices to be cleaned more often and necessitating more public services to be provided to cater for the damaged health of people. All these remedial actions use up productive resources and are costs to society at large which the perpetrator of the nuisance does not take into account unless compelled. He arranges his production according to his *private* costs only—the cost of labour, fuel and other inputs—with scant regard to the costs to the rest of society of his activities. However, the full cost of bricks to society (termed their *social cost* of production), on which the decision of how many bricks should be produced in the economy should be based, must take into account both the private cost of production and their associated external diseconomies.

Having recognised that, from society's viewpoint, a wider view of costs and benefits must be taken, the optimum quantity of any good or service will be where the marginal social cost of its production equals the marginal social benefit its consumption brings. Returning to our drainage example where the social benefit was greater than the private benefit, in order to balance the benefit to society (marginal social benefit) with the cost involved (social cost which we will assume for the present to be the same as the private costs borne by Farmer Bloggs) and so move towards the optimal use of resources from a whole-society view, it will be necessary to encourage Farmer Bloggs to invest in a greater quantity of drainage than he would otherwise do. This could be done by modifying the price system in the form of a grant on the cost of drainage. The external benefit of his action is thus brought into his decision-making process (or "internalised") by the grant which lowers his private costs, so that he equates marginal (subsidised) private costs with marginal private revenue at a higher level of drainage installation.

In the United Kingdom we see the price mechanism modified in a whole range of situations where external economies are associated with production or consumption, although there is only sufficient space here to touch on a few. The Government supports training schemes and gives grants to research establishments; part of the reason for providing grants for further education is to obtain for the rest of society the benefits of a highly educated and trained sector of the population in the form of a more rapid rate of technological development and economic growth. Grants are available to farmers willing to make holiday accommodation available in their houses; benefits accrue to the wider community, which the individual farmer does not take into consideration, through retaining employment in the countryside, preserving the appearance of the landscape for the enjoyment of townfolk and obviating the necessity for alternative and perhaps more costly Government measures to support the incomes of farmers. In the case of subsidies to rural bus services which would otherwise close down, it can be argued that the welfare of the rural community is greatly enhanced by the continued existence of the service, even if many of the benefits are not easily measurable. One such benefit might be the peace of mind given to rural non-car-owners from feeling mobile and part of the whole community rather than cut off, easing family pressure on them to move to an urban environment.

We turn next to situations where attempts are made to take external diseconomies into account, narrowing the gap between private and social costs of production. Again a few examples must suffice. In the case of the smoking brickyard chimney, where the social cost of brick production exceeded the private cost, the brick manufacturer could be compelled to take the social costs more into account by legislation forcing the brickworks to adopt methods which eliminate offensive smoke at source, such as the installation of filters on the chimneys or the use of a different fuel. Inevitably bricks would then cost more, but only the buyers of bricks would feel the direct effect, not the housewife or office worker. More costly bricks would mean that less would be demanded and hence less produced and the output would be contracted to nearer the level where marginal social cost was in line with marginal social benefit.

Another method of mitigating external diseconomies, appropriate in the case of aircraft noise, is to require the initiator to meet the cost of remedial equipment, such as double glazing of domestic houses for sound insulation. As a result, air fares have to rise above the level necessitated solely by "private" costs, but the ordinary householder who never travels by air is little affected by this. Without insulation, he has to bear a cost of air travel (noise) without enjoying its benefits of speed and convenience. This requirement that the creator of the external diseconomy has to face the cost of cancelling out his harmful effects on others is often called the "Principle that the Polluter Pays". Another method is to set up the necessary legislation to enable compensation to be exacted for external diseconomies suffered—a farmer who knows he can be made to pay for damage done to down-stream fisheries or fined heavily for polluting rivers will be cautious in his use of herbicides. Private and social costs are, therefore, kept better in line although the legislative system itself will not be costless. Publicly financed advertising is used to dissuade drunken drivers from taking to the road and imposing external costs on the rest of society. Certain countries will not allow visitors to enter unless vaccinated against infectious diseases no longer endemic in the country, since the social cost of readmitting the disease is viewed as being enormous and far greater than the benefit which any visitor who refused to co-operate could bring. Action in all the above cases is more likely if the source of the external diseconomy can be pointed to precisely. It is much easier to take steps against a chemical factory which is polluting a river (a point-source) than it is to combat background noise arising from an increased amount of traffic on city roads.

The production and consumption of many goods and services will have both external economies and external diseconomies associated with them. For example, the country bus service which bestows an external economy by reducing loneliness may also impose an external diseconomy by damaging country lanes developed for less weighty vehicles; neither externality is taken into account by the bus operator when fixing fares and the frequency of his service. The farmer who spoils the view from a cottage by erecting an ugly building in front of it may also be providing a windbreak which benefits the cottage's garden and lowers its heating bills.

The general picture of private and social costs and benefits, the solution produced by the market based solely on private costs and benefits, and the effects of modifying this solution in line with the wider social implications are illustrated in Fig. 7.2. This is a diagrammatic generality and is not based on a particular case. The extent to which external diseconomies and economies are taken into account (i.e. are internalised in the cost-benefit decision) will depend on society's awareness of them, their seriousness, the practicality and administrative cost of doing something about them, and the political will to act. It is not necessary for precise estimates of externalities to be made before action can be taken; for example the potential deleterious effect of some persistent insecticides on human health is obvious enough to cause their banning from general use. In terms of Fig. 7.2 the social cost at *all* levels of use is so high that the only intersection of marginal social cost and marginal social benefit is when zero quantity is in use; furthermore, a complete ban may be the only enforceable method of control.

Society is becoming increasingly aware of the externalities of production and consumption. The action of individuals and firms is now more correctly seen as being inter-related in ways which are not reflected through market prices. Agriculture's impact on the environment rose to political prominence in the 1980s with the realisation that farming created external costs, such as pollution of water courses, reduction in wildlife diversity, changes in landscape appearance and less easy access to land for recreation. As will be demonstrated in Chapter 10, Government agricultural policy now takes externalities into account. Many major changes to the economy, such as the siting of an airport, would not now be undertaken unless some attempt had been made at a wide-ranging cost-benefit analysis. This would not be restricted to private costs and private benefits, such as the cost of the land, buildings and staff on one hand and the anticipated revenue from flights on the other, but would also take into account the externalities of the project which a private airport operator would not consider, such as the implications of more traffic on the approach roads for the people living

FIG. 7.2   *Divergences Between Private and Social
Marginal Costs and Benefits Arising from the
Production and Consumption of a Good*

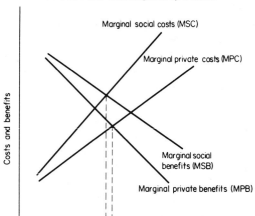

Notes: (1) In this example the socially-optimal quantity of the good is less than that
which would result from the consideration of private costs and benefits only.
However, the socially optimal quantity under other sets of conditions can be
greater, for example, if in the above diagram the MSC line were very close to
the MPC line.

  (2) It is possible to envisage situations where the MSB lies below MPB, e.g.
where external diseconomies of consumption are involved, or where MSC lies
below MPC, e.g. where unemployed productive resources exist so that more
of the good can be produced without sacrificing the output of any other good.

there, the noise disturbance to householders from aircraft, loss of
earnings and disruptions to the farm workers displaced and the deterioration of the countryside's appearance on the cost side, and the
reduction of congestion at existing airports and employment generated
locally among the benefits. On both sides of the balance there will be
elements which are difficult to quantify, frequently of the environmental
type, but they may be some of the most influential in the decision on
whether the project goes ahead.

Supporters of the centrally-planned economic system would suggest that it is superior to the capitalist free-market economy in taking into account the full costs of production and consumption. Decisions for society can be made on the basis of social costs and social benefits. Prices are not the result of market forces but are the result of planning, so that from the outset they can be set at levels which reflect not only "private" costs and benefits, but also any external diseconomies or economies. Such economic systems, however, still face the twin problems of identifying and assessing such externalities.

Public Goods

There exists a group of goods called "public" or "social" goods where external economies are particularly relevant. The services provided by the police force and national defence are of this type. If one man decided not to contribute to their cost, he would still benefit from the peace and security which the existence of the armed forces provides. The spending by the rest of the community on these services bestows an external economy on him and he cannot be excluded from benefitting, i.e. public goods are non-excludable. Another feature of public goods is that one person's benefit does not diminish other people's benefit from the same good. For example, the protection the British army affords me by deterring the forces of a potential aggressor does not stop my neighbour enjoying the same protection—we are non-rivals in the consumption of defence. Public goods, then, are characterised by *non-excludability* and *non-rivalness*. In contrast, a good such as a tractor is both excludable and rival. Legal possession implies that the services of the tractor belong to its owner and everyone else can be exluded from using them (unless, of course, it is such a handsome machine that it can be considered as providing visual pleasure, and it is difficult to stop admiring glances from the rest of the neighbourhood). Tractors also exhibit rivalness since, if a farmer buys a tractor other farmers are denied the possibility of acquiring it, and the more tractors one farmer buys the less are available for the other farmers; all the farmers are rivals for the tractors.

The market mechanism is at its best when handling goods which fall into the excludable-rival category, but it cannot be relied on to provide the socially-optional quantity of public goods, i.e. that quantity where their Marginal Social Cost just balances the Marginal Social Benefit they bring. By itself the market mechanism would result in too little public goods being provided. For example, although most people would agree that some defence force is necessary and might expresss the willingness to contribute to defence costs as long as everyone else paid their share, the unhindered market system would not result in an appropriate amount of defence being provided. A formal proof of this statement's validity could be made, but it is sufficient here simply to point out that the benefits from defence apply to everyone, and people who fail to contribute to defence costs cannot be excluded from benefitting from the spending by others. Consequently there will always be a tendency for individuals to opt out of paying for this public good and become "free riders", thereby reducing the amount spent on defence below the optimum level. Society finds it necessary for the Government to finance such services from tax revenue which is collected on a compulsory basis from individual members of the community.*

Similarly, although all UK farmers benefit from the public control over the health of animals which are imported into this country, it is most unlikely that an adequate inspectorate would be provided by private enterprise with the necessary powers to refuse entry to unhealthy stock. Human health provides another important example, treatment against many infectious diseases is only effective if everyone agrees to take part in its control, which in effect means action organised by the State.

Public finance of a service should be distinguished from public production. For example, while the control of foot-and-mouth disease by slaughter can only be effectively financed as a public policy, there is no reason why the slaughterers have to be Government employees; private vets could be contracted to do the job and paid by the State. On the other hand, there are strong arguments against entrusting the defence of this country to a private enterprise army hired by the State but under the control of some mercenary commander who might not have unswerving

---

* Two further categories of goods are worth mentioning, rival non-excludable goods and non-rival excludable goods. Honey bees are an example of the former, as the owner of the bees is unable to exclude his neighbours from the pollinating services of the insects. A seat in a less-than-full train demonstrates the latter; the ticket-holder can exclude other people from using his seat, but his travelling does not prevent other people being transported over the same journey, i.e. they are non-rivals. The market system finds difficulty in handling both of these categories.

loyalty to his "customer" if offered a bigger fee by a belligerent foreign power.

Imperfect Knowledge by Consumers

At least part of the reason why the State intervenes in two important sectors of consumption—health and education—is because it feels that individuals do not have the necessary information on which to make decisions on the amount of these commodities they wish to consume. Leaving aside the arguments for intervention in the health treatment market on the grounds of externalities and that health treatment is in part a public good, if experts can show that non-experts are under-consuming health care (or over-consuming dangerous commodities) then the State may wish to intervene to correct the level of consumption. Cigarettes are a suitable example; the link between their consumption and subsequent ill-health seems inadequately appreciated by smokers and the Government in the United Kingdom attempts to correct this by advertising the dangers involved, hence trying to change consumers' tastes and buying patterns, and raising their price through taxation so that fewer are bought.

As with many Government measures, taxes on tobacco attempt to serve more than one purpose, in this case both the consumption-changing function and the traditional role of collecting revenue for other services provided collectively. Similarly, with free health care and education financed by taxation, some income redistribution from the rich to the poor is simultaneously achieved since, by and large, the wealthy pay a bigger proportion of total tax than they receive in proportion of total benefit from these services. Some people would argue, however, that the form this provision takes, with the National Health Service and State schools run as public institutions, is a relatively inefficient method of realising the Government's community health, education and income redistribution intentions. A more effective system might be simply to give the less well-off members of the community income supplements financed by taxation, accompany this with a Government-financed advertising campaign aimed at improving public knowledge of health measures and the benefits of education, and allow individuals to spend their enhanced incomes as they wish. According to its supporters, such an alternative approach would be devoid of the heavy and paternalistic State direction of "what is good for them" which they maintain is built into the present mechanism.

Other Imperfections in the Market System

In Chapter 6 it was stated that the market for factors of production was by no means frictionless. Labour which is unemployed in one region does not flow swiftly and easily to an area in which jobs may be available. Information on job opportunities does not necessarily reach those people who would like it. When labour does move, the non-monetary costs of breaking social connections and of re-establishment in a new community may be by no means insignificant. The State has become involved in the labour market through trying to remove inefficiencies in the job information system by establishing and running job centres, and attempts to direct industry by financial incentives or planning control to locations where labour and other resources are available.

It was also noted in Chapter 6 that wages do not fall easily in times of unemployment to expand the numbers of workers taken into industry. Labour markets do not work in quite the same way as markets for, say, potatoes and wheat, in which prices respond quickly to changes in supply and demand, falling as well as rising. The labour market is apparently far more rigid, influenced by trade union strength, employment legislation, long-standing differentials and a sense of what is fair. In consequence, during a time of economic depression governments which are committed to maintaining a low level of unemployment will need to resort to more active policies to achieve their goal. These policies will be concerned with the level of economic activity in the economy, because this bears a close relationship to the level of employment. Rather than just acting as a redistributor of the income generated within the economy from one sector to another (as with unemployment benefit or pensions paid from taxation), or as a provider of public goods again financed by taxes, or as a modifier of the pattern of production so that it matches more closely social costs and social benefits, the Government is attempting to influence the *level* of national production. The factors determining the level of economic activity in a country and the manner in which a goverment can influence them to achieve its goals, not only that of achieving high employment but also others such as controlling inflation and stimulating economic growth, are covered in the next chapter.

Exercise on Material in Chapter 7

7.1 Which of the following might be reasons why the Government wished to involve itself in the economy?
 (i) The presence of externalities in production.
 (ii) Monopoly among suppliers.
 (iii) Imperfect knowledge among consumers.
 (iv) Inadequate provision of public goods.
 (v) Inequitable income distribution.

7.2 Which are examples of external *dis*economies?
 (i) Nitrogen pollution of rivers by farmers.
 (ii) Economies of scale in marketing farm crops.
 (iii) Pretty orchards for tourists in the countryside.
 (iv) High wages of auctioneers in a buoyant land market.
 (v) Spray drift which controls insect pests in neighbouring gardens.

7.3 Strike out the inappropriate words:
 From society's viewpoint the optimum amount of a commodity to produce is when marginal *private/social* cost is equated with marginal *private/social* benefit.

7.4 Bricks cost £x per ton to make in terms of clay, energy and the other costs faced by the brickworks. In a week y tons of bricks are made. People living near the brickworks are involved in £z expenditure per week in additional cleaning of their houses to compensate for the emission from the brickworks chimney. Ignore other costs.
 (i) What is the private cost of bricks per ton?
 (ii) What is the total external diseconomy?
 (iii) What is the social cost of bricks per ton?

7.5 Give three ways in which an external diseconomy may be taken into consideration in the production and consumption decisions of society.
 (i) . . . .
 (ii) . . . .
 (iii) . . . .

7.6 Give three ways in which an external economy may be taken into consideration in the production and consumption decisions of society.

(i) ....

(ii) ....

(iii) ....

7.7 What are the two characteristics of a pure Public Good?

(i) ....

(ii) ....

7.8 Which of the following can be considered as Public Goods?

(i) public buses.

(ii) defence forces.

(iii) lighthouses for shipping.

(iv) the police.

(v) cleaner air.

7.9 Which of the following are mainly publicly provided (operated) and/or publicly funded in the UK in 1989?

	Publicly provided	Publicly funded
(i) NHS hospital care	yes/no	yes/no
(ii) State education	yes/no	yes/no
(iii) NHS dental care	yes/no	yes/no
(iv) British Rail (making profits)	yes/no	yes/no
(v) Royal Mail	yes/no	yes/no

CHAPTER 8

Macro-economics — the Workings of the Whole Economy

MACRO-ECONOMICS is the study of how the economy as a whole works. Up to now we have been looking at bits of the economy separately— offering explanations for how consumers behave, how individual firms organise themselves, how the demand for the resources available for production arises at the firm and industry levels etc.—but we must bear in mind that all these elements are linked together and react with each other as broad aggregates (or totals) e.g. aggregate output, which is the total output of all firms in the economy. It is important to try to understand how the aggregate economy works because many of the economic problems which have to be faced in the present century are not caused by the action of single individual consumers or firms, and the action of individuals cannot remedy them. They are not restricted to isolated fragments of the economy but are essentially collective both in cause and remedy.

Some of the basic macro-economic problems are these; the problem of maintaining the value of money to prevent the harmful effects of inflation; of ensuring that the country's productive resources are so employed (especially manpower) that as high a living standard as is attainable within the limits set by the resources available is achieved; of encouraging economic growth; of encouraging beneficial trade with other countries and simultaneously achieving stable exchange rates between national currencies. All these problems can only be tackled from a combined (or aggregate, or whole-economy) standpoint.

As was noted in the preceding chapter, the early economist Adam Smith believed that, as long as each member of society acted in his or her own business interest, the national economy would automatically take care of itself and the interests of society would also be best served.

It would be as if an "invisible hand" were directing events. Governments should therefore adopt a *laissez-faire* policy and not try to regulate the economy, since such meddling would almost certainly do more harm than good. For reasons discussed in Chapter 7 we now know that Smith's theory was not correct, unless heavily qualified, and that some Government interference with the economy on behalf of society is required if many of the objectives which society sets itself are to be attained. A good example is the prevention of widespread unemployment, such as occurred in the UK in the 1920s and 1930s, which can be achieved by Government action. However, too much Government interference, or incorrect measures, can make problems worse, and some would argue that the Government should restrain itself rather than take actions whose full implications are not understood. An added complication is that some of the goals for which Government aims are mutually incompatible—actions taken to control inflation, for example, appear often to result in a rise in the rate of unemployment. In such cases the Government has to balance its priorities.

As a first step in analysing the workings of the aggregate economy and offering explanations for macro-economic phenomena such as large scale unemployment and inflation, we can construct a simple model of the flow of income within the economy and use it to explain what determines the level of that flow.

The Circular Flow of Income

When an individual is asked the size of his annual income he will usually give an answer in money terms because he is paid in money. But really what he is implying is that his income gives him a certain purchasing power to obtain goods and services. Buying and enjoying them are what give him satisfaction and money is only used as a convenient common denominator. Similarly, at the national level the country's income is really the quantity of goods and services which are produced and consumed by its inhabitants in a given time period, usually one year. All other things being equal, an increase in the quantity of goods and services produced and consumed which had been caused, for example, by an improvement in the efficiency of manufacturing processes implies that the nation's income has gone up.

In economies where money is used, consumers use money to buy their requirements from the market which in turn receives its supplies from firms and other producing units. This applies as much to the demands by consumers for electricity which is supplied by a nationalised electricity generating industry as it does to their demand for shoes which is supplied via high street shops by privately-owned shoe factories. There is a flow of goods and sevices from producers to consumers and an opposite money flow in payment from consumers to producers.

But where do consumers get their income from? They earn it by working in factories, offices, schools, etc.; they are selling the service of their labour (which is a factor of production) in exchange for wages and salaries. In a basically capitalist economy, such as the UK, individuals may also own land which they can rent out to producers, or capital which they can lend directly or through the banks in return for interest. Factors of production then (labour, land and capital*) are put at the disposal of producers in exchange for payment to their owners—the consumers. There is thus a flow of payments from producers to consumers, which consumers then pay back to producers as they buy shoes, food, electricity etc., forming a circular flow. A beneficial arrangement exists; the consumer sells the services of the factors of production which he controls and ends up with goods and services which he wants. Money enables the process to take place far more easily than would be possible under a primitive barter economy. This circular flow is illustrated in Fig. 8.1.

In the real world the circular flow is not so simple. As an illustration, imagine what would happen if consumers, instead of spending all their incomes on goods and services, put part away in tin boxes under their beds "for a rainy day". Unsold goods would tend to pile up in shops and in manufacturers' warehouses because more was being produced than was being demanded with the consumers' reduced spending. Because of their lower sales producers would soon cut back on their output and would no longer require to employ as much labour and other factors. In short, the whole circular flow would shrink, fewer goods and services would be produced and consumed (that is, the national income would fall) and unemployment of productive resources would arise.

---

* Entrepreneurship has been omitted purely for simplicity; entrepreneurs are also consumers and they "sell" their skills to themselves, but without the need for money as a medium.

FIG. 8.1   *Simple Circular Flow of Income*

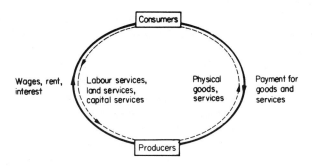

Saving is a leak, a *withdrawal*, from the circular flow. Other withdrawals, which have similar effects in running down the circular flow are: payments by consumers to foreigners for goods bought from abroad, such as imported food and the purchase by producers of foreign-made equipment and raw materials from abroad; payment by producers to the foreign owners of capital invested in the UK, such as interest paid by the Ford organisation to its American owners; and taxes taken by the State from both consumers (e.g. income tax) and producers (e.g. corporation tax).

FIG. 8.2   *An Elaborated Circular Flow of Income*

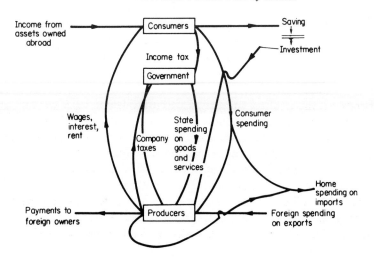

Fortunately a parallel set of *injections* exists to counteract the effect of the withdrawals, tending to increase the size of the circular flow and hence raising national income. Investment in new factories and machinery is undertaken by producers and financed partly by borrowing from individuals and from banks; goods which are exported bring in money to producers from abroad; inhabitants in this country owning property and capital abroad receive annual rent or interest payments; and the Government spends money on buying goods such as defence equipment and drugs for the National Health Service and on employing civil servants. All these injections are payments into the circular flow but originating outside it. These injections and withdrawals are combined in Fig. 8.2.

Over a given time period as long as the magnitude of total withdrawals is the same as total injections there will not be any reason for producers to expand or contract their level of output of goods and services—output will remain the same. What happens, however, if the two do *not* balance? If withdrawals exceed injections does the production of goods contract indefinitely? The answer we observe from the world about us is, obviously, no.

The reason for this is that the size of the withdrawals which people are willing to make is a function of the level of National Income, while plans for injections are very largely (and, for the sake of simplicity, taken here to be completely) independent of National Income. Individuals will want to save part of any increase in income they receive, so that total intended saving (a withdrawal) will be greater at higher levels of National Income; but the size of investments (injection) businessmen are willing to make depends not on the level of National Income but on their confidence in the future, the rate of interest on borrowed funds, etc. As people's incomes rise they want to spend part on imported goods, so the level of imports (a withdrawal) rises; but the amount foreigners buy of our products (injection), while being dependent on the size of *their* incomes, is not affected by the size of *our* incomes. Similarly, taxation systems are normally designed so that with higher national incomes more is raised in taxes (withdrawal); but the Government can spend less than it raises (and produce a budget surplus), the same, or more (by borrowing) depending on what it considers the most appropriate level of injection for the economy at the time. Overall, the intended level of withdrawals rises with national income, but the intended level of injections is independent of it. This is shown in Fig. 8.3 and from this diagram it can be seen that there is only one level of National Income at which planned (or intended or desired) injections are of the same size as planned (or intended or desired) withdrawals. This is the equilibrium level of National Income and, as long as the intentions of people and institutions who make withdrawals and injections do not alter (and ignoring for the present external influences like droughts or wars and the longer-term effect of Economic Growth), there is in the short run no reason why National Income should change once it has become established at its equilibrium level. Furthermore, the actual amounts withdrawn and injected will coincide with the planned amounts and people will have no cause to revise their intentions.

Imagine, however, what might happen if people plan to inject less into the circular flow of income. For simplicity, let us assume that savings are the only withdrawal and investment is the only injection, i.e. we are assuming that there is no Government and no international trade. Take a situation where investors in new machinery and factories for some reason lose confidence in the future and decide to invest less than hitherto. The "injection" line of Fig. 8.3 will fall, cutting the "withdrawal" line at a lower level of National Income. In practical terms this will mean that the demand for new machines etc. will fall because investors will want less of them, less will be produced and fewer men needed to produce them, or less overtime worked. Less money will in total be in people's pockets, so they will want to save less, and the firms which produce machines will be less able to put money aside for later expansion. The general level of output and income falls to a level where planned savings (withdrawal) again equals planned investment (injection). As long as investors maintain this lower level of planned injection there is no reason why National Income should rise

FIG. 8.3   *The Equilibrium Level of National Income*

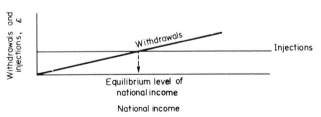

FIG. 8.4   *The Effects of Changing Levels
of Saving and Investment Intentions
on National Income*

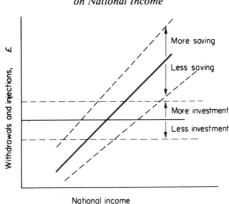

from its new low level. There is little reason in the short run why any pool of unemployed labour which might result from this lower level of total production should disappear; history has shown that wages do not fall in times of unemployment by anything like the amount necessary to mop up the pool of available labour by making the factor cheaper and hence more attractive to businessmen.

If, on the other hand, investors are prompted by premonitions of a bright and confident future for the country to raise the level of their spending, the "injection" line of Fig. 8.3 will rise, intersecting the withdrawal line at a higher level of National Income. Demand for machinery and other capital goods will be strong and their manufacturers will try to increase output by taking on more men. Output of goods will rise. In total the income of workers will increase (because more will be employed and earnings higher) so they will be able to save more and, together with the savings of firms, the total level of saving will continue to rise until it reaches the same level as investment when there will be no further tendency to rise. National Income will have risen.

National Income will thus rise if the intended level of investment injection increases, and fall if it falls. Similar effects would occur if people's propensity to save were to change. If at each income level they decided to save more, the savings-withdrawal line would tend to rise and a lower equilibrium National Income would result; conversely, if they decided to save less, National Income would rise (see Fig. 8.4). If a government is attempting to control the size of the National Income, clearly it should be aware of what is happening to the intended levels of saving and investment in the economy. Why it would be interested in doing so and the actions it might subsequently take will be returned to later.

The Multiplier

So far we have pictured the effect of changing levels of investment injection only on the activity of the capital goods industry—the producer of factories and factory machinery—and its workers. In real life the effects of injections into the economy spread much wider; if more workers are employed in the capital goods industry, part of their extra earnings will be saved (as already mentioned), part taken away in taxation and part spent on goods from abroad, but much of it will be spent on goods and services from other sectors of the national economy—clothes, cars etc., so the demand in these consumer industries will rise and production will be expanded to meet it. In turn, workers in these expanding industries will have more in their pockets, some of which will be spent on domestically-produced consumer goods. Firms producing consumer goods will probably require more plant and machinery, reinforcing the original rise in demand for capital goods. Thus the effect of an initial amount of spending in one sector of the economy spreads throughout the whole economy, rather as a stone dropped in a pond causes ripples to spread over the whole surface. Also, like ripples, the effects of the spending diminish with distance from the initial spending because, at each stage of income which the initial spending generates, a proportion is withdrawn and not passed on. Spending an extra £1m on new machine tools (injection spending) could cause *total* increased spending to add up to £4m because of the ripple effect, of which £1m will correspond to the initial spending on machine tools and £3m to the subsequent rounds of generated spending. The relationship between the initial spending and the total amount of spending is called the "multiplier".

$$\text{Multiplier} = \frac{\text{Total spending generated (including initial increase in spending)}}{\text{Initial increase in spending}}$$

In our example the multiplier would be 4. The size of this multiplier can be shown to be related to the fraction of each additional increase in spending which is *not* passed on in a second round of domestic spending because of saving, taxation or the purchase of foreign goods. This fraction of the marginal £ which is not passed on is termed the *Marginal Propensity to Withdraw (MPW)*.

The multiplier is numerically equal to the reciprocal of the MPW. For example, if in passing round the circular flow of income one quarter of each additional £ of spending would be withdrawn through saving, taxation or spending on imports etc. (i.e. the MPW is 1/4), then the multiplier will be 4. Similarly, if only 1/6 is withdrawn, the multiplier will be 6.

In the explanation of the equilibrium level of National Income given on pages 240-42 a very simple model of the economy was taken in which saving was assumed to be the only withdrawal, and investment was assumed to be the only injection. In this case the Marginal Propensity to Save (MPS) is what determines the size of the multiplier, the MPS being the fraction of each additional £ of income which is saved (the only withdrawal in the model). A MPS of 1/3 will give a multiplier of 3. The fraction which is not saved must be spent on consumption goods, and forms the Marginal Propensity to Consume (MPC). The MPS and MPC must therefore sum to one, i.e. MPS + MPC = 1.*

The existence of the multiplier effect is of great relevance to Government policy. For example, if a pool of, say, 500,000 men are unemployed in the country and the Government decides that, as a matter of policy they must all be given jobs, it can achieve this by placing orders to build

---

* Marginal Propensities must be carefully distinguished from Average Propensities. The *Average Propensity to Save* (APS) is the proportion of a person's total income which is saved, i.e. APS

$$= \frac{\text{Amount saved}}{\text{Total Income}}$$

Similarly the proportion of total income which is spent is the *Average Propensity to Consume* (APC)

$$= \frac{\text{Amount spent on consumption}}{\text{Total Income}}$$

Since what is not saved must be spent on consumption, the Average Propensity to Save plus the Average Propensity to Consume must also sum to one, i.e. APS + APC = 1.

roads with construction firms who will employ more men. However, it is not necessary to place sufficient new road contracts to absorb *all* the available labour. The Government knows that much of the money it pays to construction firms will be spent by their employees and shareholders and generate expansion of other sectors of the economy which will in turn demand more labour. All the Government has to do is to place sufficient road orders, which may themselves only need 100,000 men, to generate a *total* demand for 500,000 men. If it goes beyond this a shortage of labour will eventually result. Another example might be where the Government encourages a firm, looking for a site to establish a new factory, to set up in a region where incomes and living standards are lagging behind the rest of the country. It knows that the wages paid by the factory will generate spending and incomes right through that region far greater than the initial amount paid out in wages.

The multiplier effect has often been described as a two-edged sword. This is because, just as stimulating increases in spending (injections) can spread throughout the economy, so can depressing withdrawals. If, for example, a large company were to reduce the amount it normally invested in new machinery, firms making machines would face reduced demand and would lay off workers who would then cut down their spending on consumer goods, and so on. Total output (and incomes) of the commodity would fall, as might well the level of employment. In such circumstances the Government would be under political presssure to counteract the tendency for the economy to run down by spending more itself.

The Level of Aggregate Demand*

An alternative way of analysing the circular flow of income is to consider the level of aggregate demand, alternatively called aggregate expenditure. Reference back to Fig. 8.2 shows that the demand for goods and services comes to producers from three sources—consumer spending (C), spending on capital goods in the form of investment (I), and spending by the Government (G). For simplicity we will assume that spendings on exports and imports equal each other and can be

---

* The term "aggregate" demand is used to mean the combined demands of the several sectors of the economy.

Fig. 8.5   *The Range of Possible Equilibrium*
Combination of National Income
and Expenditure

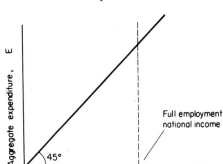

ignored at this stage. As investment and government expenditure are both injections, it is possible to think of aggregate demand as consumer expenditure (C) + injections (I + G). As long as aggregate demand is just sufficient to absorb the quantity of goods and services that the nation is producing, the economy will be in equilibrium. Put another way, National Income, which is the goods and services produced in a year expressed in terms of money, will have no tendency to rise or fall if it is equalled by the value of total expenditure; if expenditure is greater or less than National Income, then National Income must rise or fall before equilibrium is restored.

Fig. 8.5 shows all the levels of National Income which are potential equilibrium combinations of income and expenditure, i.e. where income (Y) and expenditure (E) equal each other—this is simply a line at 45° between the income and expenditure axes. Any National Income on this line is maintainable indefinitely, up to the limit set by the country's productive resources. Only a certain quantity of labour, capital and the other factors of production are available at one time, and once these are all fully employed no further increase in National Income is possible, at least in the short run. Below this full-employment level of income a whole range of equilibrium National Incomes is possible, although some unemployment of resources will then occur.

The Determination of the Equilibrium Level of National Income

Expenditure within the economy comprises, as already mentioned, expenditure by consumers (C), expenditure on investment (I) and expenditure by Government (G). We have also already noted that the latter two (I + G) can, as a first approximation, be assumed to be independent of the level of National Income. Consumer expenditure is, however, not so divorced, and relates to National Income in the way shown in Fig. 8.6. As National Income rises, so does the amount consumers desire to spend, but not at the same rate; if it did it would have the same slope as the E = Y 45° line. At incomes greater than N planned spending is less than the National Income, and so some is saved. At levels of Y below N consumers will want to spend *more* than the National Income in order to maintain their living standards, implying that they must dig into past savings to do so. If there were no investment going on in the economy and no Government spending,* National Income would be in equilibrum at N because, at that level, intended aggregate expenditure and National Income would be equal. There would be just enough demand to take up everything that producers were putting on the market. Any attempts by producers to produce more would simply result in stock accumulating on shelves, and production would be cut back to its original equilibrium level.

Fig. 8.6 *Consumer Expenditure and National Income*

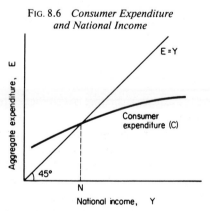

---

* An unreal situation, because without investment in capital goods, production of consumer goods would soon come to a halt.

Fig. 8.7   *The Equilibrium Level of National Income*

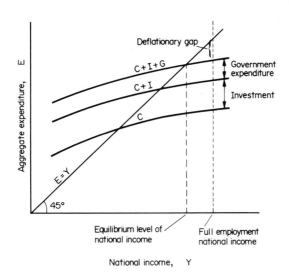

Introducing investment (I) and Government spending (G), the other two components of aggregate demand, into the model results in Fig. 8.7. Note that, if I and G are independent of Y, then adding them to consumer expenditure gives lines for total national expenditure parallel to but higher than the original curve of consumer expenditure. The equilibrium level of National Income is where the combined level of expenditure (C + I + G) intersects the E = Y 45° line. At this level intended aggregate expenditure just balances the value of goods and services produced so people have no reason to revise their intentions and National Income will be stable at this level.

It is worth noting that the equilibrium level of National Income shown in Fig. 8.7 could also have been prescribed by the method described earlier—the balance of intended injections with intended withdrawals. The same equilibrium income level would have resulted in either case. Only the ways of analysing the situation differ.

Policies to Control Unemployment

After the Second World War Governments in the United Kingdom accepted responsibility for preventing high levels of unemployment. This they attempted to achieve principally by encouraging National Income to rise until any serious under-utilisation of the country's labour resources was wiped out. This policy, developed from the thinking of John Maynard Keynes (1883-1946) was built on the belief that, without Government manipulation of aggregate demand, there is little reason why the equilibrium level of National Income should correspond to high or full employment levels. Keynes pointed to this as the explanation for the prolonged period of low national income and high unemployment in the late 1920s and 1930s; less enlightened economists had thought that full employment was the norm to which the economy would eventually return.

In Fig. 8.7 the position at which the C + I + G line (aggregate expenditure) cuts the 45° line indicates an equilibrium level of National Income corresponding to less than that which could be generated when all resources are fully employed. To raise the level of employment aggregate expenditure must be increased to make good the short-fall between the level of expenditure which would occur with the present C + I + G line in Fig. 8.7 at full employment and the National Income at full employment. This short-fall is termed the "Deflationary Gap". If businessmen can be encouraged to invest more, perhaps by giving them confidence in future prosperity, the I component will increase, and the C + I + G line will shift upwards. Or, which is the more likely, the Government can spend more on roads, hospitals, social services etc. increasing the G component and raising the C + I + G line, creating a higher level of equilibrium National Income corresponding with a higher employment level. They must be careful, however, not to raise C + I + G to such a level that the equilibrium National Income is beyond that which is attainable with the full employment of all available factors (see Fig. 8.8). Because this is unattainable, intended aggregate expenditure will exceed the value at existing prices of the quantity of goods and services which can be produced (i.e. National Income). This "excess demand" will result in continuing inflation as the prices of the full-employment output of goods and services are bid up, assuming that the supply of money is allowed to increase to

FIG. 8.8 *An Unattainable Level of Equilibrium National Income*

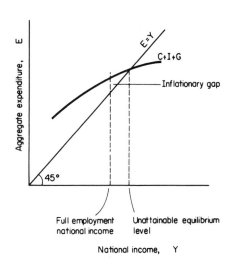

accommodate the higher prices and higher money wages which follow. In such a situation the gap between the level of expenditure at full employment and the value of output (National Income) at full employment is termed the "Inflationary Gap". Inflation will be returned to later.

Similarly, if consumers decide to increase the proportion of their incomes which they spend, the level of consumption spending in the economy will increase and the "C" curve in Fig. 8.7 will rise, taking the C + I and C + I + G curves with it. This might happen if consumers come to believe that saving is futile because anticipated inflation will greatly reduce the value of their savings. The raised C + I + G line might well produce a situation of "excess demand", that is where the equilibrium level of National Income is unattainable, itself producing the inflation which consumers fear. To counteract this, the Government could reduce its spending, lowering the G component until the Inflationary Gap disappeared and an attainable equilibrium National Income was produced.

Full Employment and the Longer Term

The economic theory presented so far in this chapter is generally accepted. But the precise workings of the national economy are by no means fully understood. In particular, there is a range of views on the extent to which governments should attempt to manipulate the economy, and on the means by which this should be done. Since the early 1970s serious doubts have been cast on the advisability of striving for full employment as its pursuit by successive administrations through demand management along Keynesian lines has been blamed for the inflation which has increasingly dogged the United Kingdom post-war economy. The argument is that, while expanding Government expenditure will eliminate unemployment and increase output to the full employment level in the short run, i.e. a boom period will be created, growth in the economy will be insufficient to sustain the necessary demand for this extra output permanently, and production and employment will fall back. Further expansion of spending by the Government will be required to sustain employment and this will produce inflation because, although the volume of money in the economy is increasing (due to the Government borrowing to finance a continual budget deficit), the quantity of goods on which it may be spent is not rising to the same extent, so their prices rise. Each time the level of output falls back, unemployment reaches a new peak.

Moreover, high levels of employment sustained by Government demand management are believed to lead to a loss of dynamic drive within industries, poor output per worker, an unwillingness to innovate, unresponsiveness to changing patterns of demand, and a failure to be competitive with suppliers from overseas. Contrary to the views of many followers of Keynes, the economy is thought to have significant self-regulating properties if only the free market were allowed to operate more in the way that industries organised themselves. These ideas are associated with the American economist Milton Friedman. They emphasise the supply side of the national economy.

The proponents of this view of the economy advocate an abandonment of a full-employment policy. They would argue that jobs are best created by industries becoming internationally competitive, and this is more likely to happen under private control than if run by the State. To this end, in the 1980s Conservative governments in the UK returned many nationalised industries to private ownership (privatisation). A hands-off approach to the economy may well mean having to accept in the 1990s levels of unemployment which would have been strongly resisted in the 1950s and 1960s, but higher levels may now be more tolerable for a variety of reasons. With the improved Social Security system the amount of overall hardship resulting from a relatively high rate of unemployment could well be no greater than that which formerly resulted from a much lower rate. With increasing awareness of job opportunities and job satisfaction, people are more careful about choosing jobs and do not necessarily take the first available, so the period during which they are prepared to be unemployed may well be greater and hence the level of unemployment in the economy may be higher. From the national viewpoint higher unemployment probably also means that labour can be switched from declining to expanding industries more easily, speeding the modernisation of the economy. In addition, if full employment policy can be shown to have been a major contributor to the causes of inflation in the 1970s and early 1980s, many members of society would be willing to accept a small degree of unemployment as preferable to rapid falls in the value of money. Supporters of this school would not claim that the Government should *never* use its spending power to stimulate National Income; there may be exceptional circumstances in which aggregate demand management may be needed. But normally the Government should restrict its role to creating the framework of stability, notably by controlling inflation, within which private industry can operate. And this can be best achieved, so they believe, not by demand management but, as we will see later, by policies which relate to the nation's supply of money and rates of interest.

Inflation

Any economy which has progressed past the stage of barter requires some common denominator against which the wide variety of goods and services it generates can be exchanged—in other words a money system. Wages are paid in the form of money and spent on items priced in money units. Great difficulty would be experienced if a farm worker were paid solely by being given a leg of a cow each week which he had to barter with the grocer and the Electricity Board. In most developed countries the monetary unit is of little or no intrinsic worth—a £1 note is just a scrap of paper—but it has exchange value because it is accepted in exchange for goods which *have* intrinsic value by all members of a country. As long as the exchange value of the monetary unit is maintained people will not feel compelled to spend all their earnings immediately because money acts as a store of value. They will save in terms of money with banks and building societies and be willing to make loans, accepting repayment in the future. However, when the value of money changes rapidly, and this usually means downwards with rapidly rising prices, money becomes less acceptable as a store of value and the smooth running of a developed economy is jeopardised.

Falling money values and rising prices erode the value of savings and reduce the living standards of those people living from savings or on fixed incomes, notably the elderly living from pensions. Borrowers benefit and lenders suffer because the nominal money size of loans, say for house purchase, does not rise with inflation, but the nominal value of houses and most incomes usually do rise. With the passing of time, repayments constitute a diminishing part of the borrower's income and become worth progressively less in terms of purchasing power to the lender. The general effect is that spending on borrowed funds is encouraged and saving discouraged, tending to exacerbate the inflation as will be shown later. Inflation, then, redistributes real income and wealth in a somewhat arbitrary and generally undesirable manner. Frequent wage claims will be made to compensate for rising prices, bringing with them industrial unrest especially if some sectors, such as those employed by nationalised industries and the public services, have wage control strictly applied to them in an attempt to halt inflation, while other sectors manage to secure larger increases. With inflation the price of exports will rise (unless an exactly compensating depreciation in the international exchange rate for the currency is allowed) making them less competitive abroad, causing an adverse movement in the balance of payments; in turn this may lead to depreciation and more expensive imports and hence higher living costs at home. Recession starting in the exporting industries can spread by the multiplier effect to other industries. This can stunt the rate of economic growth through businesses losing the confidence to invest and having to face the high interest rates which lenders will try to obtain as a way of counteracting the fall in the purchasing power of the currency. House ownership is also put beyond the aspirations of many by these high borrowing rates.

This formidable list is by no means exhaustive, yet illustrates why reasonable stability in the value of money is a highly desirable feature, almost a prerequisite, of a smooth functioning exchange economy.

The Quantity Theory of Money

Money consists not only of notes and coins but also bank deposits and loans where cheques, or some other type of purchasing power transfer such as credit cards, are involved. If a bank were to give all its customers an increased overdraft, they would spend that money in much the same way as if the bank handed them notes. However, the simple quantity theory of money states that the number of times a piece of money changes hands (called the velocity of circulation) is also vitally important and that the real quantity of money is better represented by the size of the money supply times its velocity of circulation. For example, a single £1 note changing hands in a market ten times will finance just as much total transaction (£10-worth) as ten £1 notes changing hands once. The theory begins with the tautological equation:

$$MV = PT$$

amount of Money $\times$	Velocity of $=$	general level $\times$	total number of
(bank notes +	circulation	of Prices	trade Transactions
deposits including			
overdrafts)			

The number of trade transactions (T) will depend on how many goods and services the country is producing and putting on the market.* The equation is a tautology because both sides show the same thing; the left-hand side shows the total money value of transactions over a given period while the right-hand side measures the total money value of goods sold.

From the equation several predictions can be made. Taking first the variable M, if the Government allows banks to lend more to customers, increasing aggregate demand, or if it spends more itself by borrowing in a way which increases the reserves held by banks, thereby enabling them to create further credit, the money supply (M) will increase.

---

* Sometimes y is given in the equation rather than T, y standing directly for the flow of goods and services over the given period.

Assuming that the velocity of circulation (V) is constant, as long as the quantity of goods and services (T) which the nation produces increases in step, there is no reason why the general level of prices should change, i.e. no inflation will occur. This can happen if the country has unused productive resources (as happens in a slump) which can be used to boost output. If, however, all the resources are fully employed, increasing aggregate demand and the money supply cannot be accommodated by changes in T, and P must rise instead, i.e. inflation will result as an increasing quantity of money chases a non-increasing quantity of goods. This is equivalent to the situation referred to back in Fig. 8.7 where an inflationary gap was identified. This type of inflation is termed *demand pull* because it is caused by excess demand in the economy. There is much dispute over the ability of an expansion in the money supply to *cause* a rise in demand; some economists (the "monetarists") believe that money has this active property. They point out that increasing the amount of money causes interest rates in the economy to fall. They further believe that these lower interest rates cause businessmen to invest substantially more heavily and consumers to borrow and spend more, thereby increasing the level of aggregate demand. If there is unused productive capacity in the economy, production (and National Income) will rise. If not, inflation will result. Others (the neo-Keynesians) hold the view that investment and consumer spending is not very sensitive to interest rates, being mainly influenced by other factors such as expectations of future prosperity. These deny the ability of an expanding money supply to affect the real level of economic activity, and conclude that it simply enables excess demand resulting from the intended level of Government and private spending being greater than the value (at existing prices) of what is available for purchase to be realised in higher prices. Most exponents of either view would agree, however, that a money supply which consistently increases faster than the growth in output of goods and services will be associated with rising prices.

Increases in productivity have a role to play in mitigating the inflationary effect of a rising money supply as they increase the quantity of output, from a given stock of factors, on which the rising spending power can be dissipated. Even if all productive resources are fully employed, they can often be used more efficiently by eliminating wasteful production methods, developing better types of equipment, reducing over-manning etc. In practice, however, such attempts to "grow" out of an inflationary situation in the United Kingdom have not been successful at the national level; increases in productivity have been exceeded by concurrent increases in the money supply.* On a regional basis, however, the general principle may be found working: seaside holiday areas receive large seasonal influxes of spending power which would bid up the price of food enormously if it were not for the increased supplies of food which are channelled to them; expanding towns often try to ensure an adequate supply of housing so that the arrival of big firms with staff does not result in gross inflation in house prices.

The *velocity of the circulation of money* (V) is often taken as constant, but it will be affected by people's expectation of future inflation rates. If inflation is expected to worsen, then consumers will be encouraged to spend rather than save because (a) the value of savings can only be expected to fall, (b) it is known that the money values of many durable items like houses rise with inflation, maintaining their real value, and so making their purchase highly attractive, and (c) waiting to purchase goods will often put them out of reach because of price rises. This urge to spend increases the velocity of circulation which, without further real production, aggravates the inflation that the spenders were hoping to circumvent. In inflationary times the velocity should, if possible, be slowed by giving consumers greater confidence in the future value of money and a willingness to save.

Until the mid 1960s demand-pull was the traditional explanation for inflation; such inflation could be cured, it was believed, simply by reducing the level of aggregate demand until a pool of unemployment was created. Once this was created inflationary pressure was removed because workers would not be in a strong bargaining position—if they did not accept the existing level of rewards, other men were supposed to be waiting for their jobs. Professor A. W. Phillips showed that, from the 1860s up to the 1950s, a relationship held between the level of wage and price rises and the level of unemployment, more unemployment being associated with low inflation. The general relationship was of the type shown in Fig. 8.9.

---

* As was illustrated by the policy of the Conservative Government in the UK from 1970.

Fig. 8.9 *"Phillips Curve"*
A Trade-off Relationship between
Inflation and Unemployment

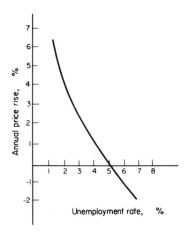

The late 1960s and the 1970s saw rates of inflation far greater than the Phillips curve, based on historical data, would lead one to expect. Both inflation *and* the level of unemployment were high whereas the simple demand-pull theory would have regarded the two as incompatible. This has given rise to a range of alternative explanations for recent inflation, most of which are *possible* but of doubtful probability as the real explanation. Most of them are of the *Cost-Push* type; these explanations point out that, under the conditions of industrial organisation and unionisation then found in the United Kingdom, increased costs of production, notably successful wage demands, were passed on as price rises which again caused further wage demands as the cost of living increased, and so the spiral continued. Cost increases which arise outside a country's economy, such as an increase in the price of imported oil, or by chance factors like a crop failure affecting the price of potatoes, are obviously unavoidable, but there seems little reason why internally-generated, self-sustaining cost-price spirals should be tolerated. The strength and "pushfulness" of Trade Unions in being able to force employers to pay higher wages by threatening or using strikes which could jeopardise the viability of firms more effectively than would the raising of prices necessary to pay the higher wages was blamed for the perpetuation of cost/price spirals, although supporting evidence that increased Union strength or strike activity was the cause of inflation is not strong. The increasing failure of real wages to keep up with expectations has been cited, as has also the insistence of workers on maintaining parity or established differentials between those industries where increases in productivity have occurred and those where none have taken place.

The most likely explanation for the inflation of the 1970s in the United Kingdom would appear to be one based on the demand-pull idea, but modified. The argument runs that, once inflation was established, it was built into any wage bargaining procedure as a basic demand, on top of which negotiations for "real" wage increases were superimposed. With an anticipated and "allowed-for" inflation established, and a Government monetary policy which permitted the money supply to expand to accommodate higher levels of costs and prices, there was no reason why inflation should slow down when the economy was close to or at the full employment level of activity. We have already shown that, while high levels of employment and high output can be stimulated in the short run by employing unused and potentially productive resources, in the longer term aggregate expenditure cannot rise faster than the sustainable increase in output of the economy (in other words, the economy's rate of growth) without producing inflation.

If the Government creates an "excess-demand" situation by creating too much expenditure and striving for an unattainable level of National Income growth and high levels of employment, inflation will accelerate from its expected base level. According to the proponents of this theory, to slow inflation aggregate expenditure must be cut, and in particular Government borrowing from the banks to cover its budget deficit spending. This creates a pool of unemployment the size of which depends on how fast inflation is to be checked. At the same time people's expectations of the future rate of inflation must be reduced. For this latter purpose a prices-and-incomes policy of restraint may be a useful tool; historically they have had little if any long-term effect but may be useful as part of a package of policy measures including a reduction in aggregate expenditure. However, they carry the danger that, because they tend to be applied more vigorously in the public sector than in privately-owned industry, they can result in industrial unrest. Concurrently the aim of monetary policy must be to slow down the growth of the money supply so that eventually it is expanding in line with expansion in national output. This is the type of policy that was followed by Conservative governments in the UK over the 1980s. Money supply was controlled and public spending cut back. High interest rates caused many less-competitive firms to close down and a great number of jobs to be lost. The power of trade unions was curtailed both by legislation (on closed-shop agreements and on strike ballots) and by their failures in disputes with determined employers (most notably the year-long miners' strike against their public-sector employer the National Coal Board). High unemployment was seen as a necessary part of restructuring industry to make it profitable and internationally competitive and to achieve an economy which was less prone to inflation.

Economic Growth

The promotion of economic growth is another of the goals of Government, although growth can go on quite happily without Government encouragement. It occurs when the amount which can be produced and consumed per head of the population increases. As will be discussed later, economic growth does not necessarily mean that the inhabitants are happier, simply that they can buy more goods and services. If we assume that the average European, because his income is higher than the average Indian's, is thereby "better off", we are making a subjective judgement about what brings happiness.

If some of the country's productive resources are unemployed it will be possible in the short run to increase National Income by increasing expenditure to take up the slack of unused potential. These short-run changes are *not* what is usually meant by economic growth.

Economic growth occurs when a country's *productive capacity per head* increases. Productive capacity is the output, measured by Gross National Product,* of a country achieved when all of its productive

---

* Gross National Product is the value of output at factor cost, with due allowances made for net property incomes from abroad, and before capital depreciation allowances have been removed.

resources are fully employed. However, in practice, this full employment qualification is often dropped and growth is taken as the change in GNP per head, after allowing for changes in the value of money. If a country's total output is expanding, but its population rising even faster, the quantity of goods and services available for each person on average will be declining. We will refer again to the race between output and population for the goal of economic development.

A country's rate of growth will depend on a variety of factors which are related to those which determine its present output per head. Principally they concern capital, labour and the state of technology. Growth will occur rapidly if the amount of *capital per head* of the population can be increased rapidly. This in turn will allow more consumer goods and services to be produced per head because each worker on average will have a greater quantity of machinery etc. to assist him. An obvious example is the man on the powerful tractor who can do so much more cultivation per day than a man with a spade. The number of men needed on farms has fallen dramatically over the last one hundred years, yet farm output has also increased; production per head is much greater because of all the capital now employed.

When the Russians wished to turn their country from its largely peasant state in the 1920s into a modern industrial nation, they embarked on two five-year plans of building up their stock of capital, first (1928-33) in the heavy industries (e.g. iron and steel) and later the lighter manufacturing ones. As was stated in the chapter on Factors of Production (Chapter 6), countries with high outputs per head (so-called developed economies) find it much easier to build up further their stocks of capital than do low output countries. Increasing capital accumulation means diverting resources from the production of consumer goods to capital goods; in a poor country all the production may consist of necessities, so the margin available for capital build-up may be very limited. Aid in the form of gifts or loans of capital from rich countries can be an important growth stimulant.

A moment's thought should raise the question of diminishing returns. Given a population of a certain size and a stock of natural resources, is not capital subject to diminishing returns, so that each successive unit of capital which is added to a country's stock of assets generates progressively less extra output? This is illustrated in Fig. 8.10 where the Marginal Product of Capital (MPC) is shown as falling as

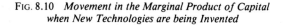

FIG. 8.10   *Movement in the Marginal Product of Capital
when New Technologies are being Invented*

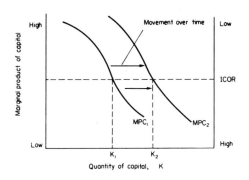

Notes: (1)   The Marginal Product of Capital is the addition to output caused by the marginal (additional) unit of capital, other inputs held constant.

$$\text{MPC} \ = \ \frac{\text{Change in quantity of output}}{\text{Change in quantity of capital}}$$

MPC is inversely related to the ICOR (see 2 below)

(2)   The term ICOR (Incremental Capital/Output Ratio) is often used in a growth context. As its title implies it is represented by a formula

$$\text{ICOR} \ = \ \frac{\text{Change in quantity of capital}}{\text{Change in quantity of output}}$$

(3)   When funds become available to be spent on capital goods the most attractive (implying the most productive) investment opportunities will be used up first, then the less attractive and so on.

(4)   In an economy where no invention or innovation took place a build-up of capital would cause the MPC to fall and the ICOR to rise in a way described by the MPC_1 curve in Fig. 8.10.

(5)   In an economy which does experience technical innovation the MPC curve will be shifted rightwards. If the stock of capital increases in a given time period from K_1 to K_2 the realised Marginal Product of Capital will not fall and the ICOR will not rise as long as technical innovation simultaneously shifts the MPC curve from MPC_1 to MPC_2.

(6)   ICORs are difficult to measure and ACORs (Average Capital/Output Ratios) are often substituted.

$$\text{ACOR} = \frac{\text{Total quantity of capital used}}{\text{Total quantity of output}}$$

ACORs bear a relationship to ICORs in rather the same way as Average Costs relate to Marginal Costs. Because ACORs have apparently not risen in the USA in the 20th century it is thereby assumed that ICORs have not risen either.

the capital stock is built up. And could not the situation arise where more capital added nothing to production, or even reduced it? While such a situation is possible, it appears that, for the richer countries, new techniques of production and new products are discovered at about the same rate as extra capital becomes available. This has the effect of shifting the MPC curve to the right (Fig. 8.10) so that marginal blocks of capital continue *to increase* national output by about the same extent.

Clearly another important determinant of the rate of economic growth is the rate of *invention and innovation*, both that undertaken by private firms and that supported by the State. A good example of an invention is the desk-top computer which enables so many business operations to be performed more rapidly than hitherto, and hence makes the work-force more productive. New plastics, improved alloys, more fuel-efficient aircraft etc. all provide new investment opportunities and permit greater production, while new consumer goods (videotape recorders, hi-fi) provide an ever-widening range in which the entrepreneur can exercise his production abilities. In the United Kingdom the State actively supports research and development—in the agricultural context the Agricultural and Food Research Council funds research. Copying good ideas from abroad is an age-old practice. In the period following World War II the Russians and Japanese were particularly successful in aiding their own growth by "borrowing" ideas from the West; this was readily observable in motor cars, where their products tended to resemble out-of-date American and European models. However, this "borrowing" of ideas is increasingly difficult once the gap between the "borrowing" and the "lending" countries has narrowed. Furthermore, the "borrowed" techniques may well be more appropriate to situations where labour is scarce than where pools of unemployed labour already exist, as is the case in many less developed countries.

The *quality of the human capital* also bears on growth. Spending money on education and training is at least in part an investment—scarce resources must be devoted to it and an educated and trained labour force can be more skilful, more productive and more adaptable to changing conditions. Literacy certainly aids productivity, but many would argue that certain sorts of esoteric higher education add nothing to a country's ability to produce and little or nothing to the quality of life of the individual or the community. Far from promoting growth, an education which encourages students to question growth's desirability may inhibit it.

Changes in the *quantity* of labour (meaning the size of population) have an important relationship with growth, as has already been mentioned. If the productive capacity of a country is increasing fast, yet population is increasing faster, growth in the sense of an expansion of productive capacity *per head* will not occur. Rapidly expanding populations are a major problem of poor countries. On the other hand, increases in the size of the population can contribute to growth under some circumstances. In Fig. 8.11 a production function is shown, with labour taken as the variable input, all others being held constant. The standard of living can be represented by the average product curve. It will then be seen that, with low labour usage, increasing quantities of labour will raise average product. Such a country could well be labelled "under-populated". An example might have been Australia early this century. Increasing the labour force, however, causes output per head to reach a peak, beyond which living standards decline with further increases in the labour force because of diminishing returns. This was the sort of thing the early British economist Thomas Malthus (1776-1834) foresaw in his predictions of what would happen to the supply of food with a rapidly rising population; because of diminishing returns, food production per head would decline as the population increased until only sufficient to maintain the population was available. If the population were to go beyond this, he believed that it would be cut back automatically by famine, disease and other miseries.* With the rapid increases in population which were occurring in the UK at the time of his writing such catastrophes seemed unavoidable. Fortunately, the unforeseen development of North America and the British Empire as suppliers of food and the possibilities of trade meant that the UK had both a larger population *and* higher living standards in 1900 than in 1800. New production techniques raised (and continue to raise) the average product curve to offset the effect of falling average product which might have been expected with a rising population. Increasing numbers are no longer a matter of serious concern to developed economies, but poor countries still face problems of population expansion and the predictions of Malthus are by no means irrelevant to them.

* Thomas Malthus (1776-1834) brought out the first edition of his "Essay on the Principles of Population as it affects the Future Improvement of Society" in 1798.

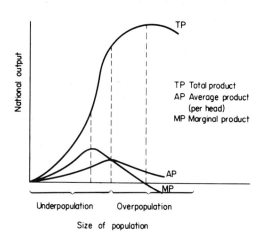

FIG. 8.11 *The Relationship between Population Growth and National Output* (assuming a strong positive correlation between population and work force sizes)

The types of *economic, social and legal institutions* with which a country is equipped are further influences on its rate of growth. A centrally planned socialist economic system has shown itself to be capable of rapid growth (USSR, China etc.) because productive resources can be directed towards the build-up of capital more readily than in capitalist economies, although this does not preclude the possibility of remarkable capitalist growth rates, as exhibited in the post-War recovery of West Germany and Japan. Agricultural land tenure is often a barrier to growth where land is fragmented into many small units, making mechanisation impractical. Landlords, often absentees, who are content with a low rent because of the large areas they own and who operate restrictive policies on their tenants are a hindrance to the achievement of higher productivity. Economic growth has often necessitated a major revision of the land tenure system. Religion has on occasion been quoted as a hindrance to growth, especially where it is mixed with social conventions. Religions which place little emphasis on materialism in this life or which restrict the availability of female labour by stressing the home-making role of the mother are not likely to encourage growth.

International trade can speed the role of growth by allowing specialisation and exchange, removing the necessity for a country to develop all its lines of production. For example, growth in Britain in the 19th century was facilitated by its ability to import food from abroad to feed its rising and increasingly urban population. In exchange it could trade exports of manufactured goods which it was best suited to produce and which were in demand by the food-exporting countries. Growth requires increasing a country's stock of capital, and trade has a role to play here. Accumulation at home may be a slow and difficult process, but trade can provide the means by which capital can be borrowed from abroad in exchange for a future flow of exports. The size of the repayments must be viewed in light of the extent to which the loans allow the country to produce and export more. For example, loans from the USA aided post-war recovery in Britain and had to be repaid by currency earned through exports. The less-developed countries are in the unfortunate position that many of their exports are agricultural and are not in high and increasing demand by richer countries. This limits their opportunity for growth emanating from trade in these commodities.

In developed economies sustained growth seems primarily constrained by the rate of build-up of capital (both investment in goods and human capital) and the rate of invention and innovation. The less-developed economies frequently face the additional hindrances of unhelpful legal, religious and social structures and institutions, poor nutrition, poor roads and communications, crude banking systems, high levels of unemployment which may be aggravated by copying labour-saving but capital-demanding production techniques from the rich countries, a bleak trading position, and often few natural resources which can be exploited for industrial development. Furthermore, the gap between slow-growing poor countries and faster-growing rich countries is an ever-widening one. Many people, particularly in the poor countries, find the disparities in material living standards increasingly objectionable as the greater ease of communication makes their existence more obvious.

The Costs of Economic Growth

Not everyone would agree that economic growth is desirable. While the poor usually strive for increased possessions and consumption of more goods and services, the lure of materialism has palled for many in the developed countries. Personal satisfaction is only in part derived from easily identifiable goods and services and a major source is the heterogeneous group comprising job satisfaction, stability in society, absence of excessive noise and pollution, and many others which are contained in the diffuse term "quality of life". Attaining higher material standards of living through economic growth imposes costs which, upon reflection, at least partly offset and may completely nullify increases in satisfaction gained from having more clothes, housing, consumer durables etc.

Growth derived from increasing a country's stock of capital can be more costly than is immediately apparent. Capital build-up involves the diversion of productive resources from making consumer goods to making producer goods, implying a sacrifice of living standards in the short run in return for higher standards in the future. This sacrifice is a cost of promoting growth. While consumption may return fairly soon to its original level, the net advantage of growth is only enjoyed after sufficient extra consumption goods have been produced to offset the original sacrifice (see Fig. 8.12). If less value is ascribed to goods which

FIG. 8.12   *Losses and Gains in Consumption over Time when Growth is Pursued through Capital Accumulation*

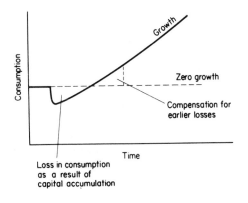

Note: A similar diagram could have been given when a growth rate of say 2 per cent is accelerated to a faster growth rate by the accumulation of capital.

can only be enjoyed many years hence than on goods which are available now (i.e. a process of discounting is employed) the attractiveness of growth derived from capital accumulation diminishes further. As most countries encourage investment as a a means of promoting growth, although probably it is only responsible for about half of the total in developed countries, we must assume that on balance such short-term sacrifices are considered worthwhile in the national view.

Higher levels of national production, it is alleged, imply costs in the form of a faster pace of life, more heart attacks, more stress and frustration, more suicides and a less-caring society. Externalities of production such as air and water pollution become of increasing importance and an increasing proportion of the nation's resources has to be used in their control. Because of the difficulty of dealing with the less tangible forms of external diseconomy such as ugliness, the quality of the landscape is frequently thought of as deteriorating through the housing and industrial development which accompany growth, despite attempts at planning control. High technology societies, as well as permitting increasing amounts of leisure time, run the risk of reducing some sections of the population to becoming machine minders with little interest in the product of their labours. Added to this, the increasing amount of mechanisation is likely to reduce the number of unskilled jobs which are available. Rapid technological advances make old skills obsolete and necessitate retraining if their owners are not to become redundant. Consumers are asked to make spurious choices between brands of goods, like soap powders, where contents are identical and only packeting differs. At the opposite pole adequate products are withdrawn from the market when new models are introduced, reducing in this case the ability of consumers to choose between the familiar and the "improved" version. Advertising seems set on convincing people that only material goods are important, that they are dissatisfied with what they have and lures them into buying more in the hope of satisfying themselves.

In the United Kingdom economic growth has brought changes in farming practice and the appearance of the countryside which many observers find objectionable from an aesthetic or ethical viewpoint. A decline in the number of farmworkers (and to a lesser extent the declining number of farmers) has been accompanied by increased mechanisation, field enlargement with the removal of hedgerows, and the greater use of herbicides, pesticides and fertilizers. Changes in the size and the composition of the rural community have taken place (by no means all being detrimental) and problems have been experienced with services such as rural education and public transport. For a detailed description of the changing face of UK agriculture and of its contribution to the growth of the national economy, reference should be made to one of the recommended texts listed at the end of this book. At this stage it is simply necessary to point out that increased levels of production either by the whole economy or by the farming industry alone should not automatically be welcomed without also examining their wider consequences.

The critics of economic growth are not short of ammunition with which to bombard the policies of governments aimed at its promotion. Their salvos have not succeeded in halting growth, nor is this their considered aim in most cases. They have achieved considerable success, however, in dragging before public attention the fact that satisfaction is not derived solely from the fruits of production—goods and services— and that measuring changes in the standard of living by ascribing money values omits many of the factors which determine the quality of life. The economist has to take into account that most changes in the economy—such as the building of a new airport—have implications for many sectors of society. Some implications will be beneficial, others deleterious. Only some are readily quantifiable and an assessment of the desirability of growth, or indeed any change in society, must in the last resort be a subjective judgement.

This section has emphasised the interrelationships of individuals and firms, both singly and collectively. Such interactions are not confined to within the national boundary and a natural extension is the notion of international trade, considered next.

Exercise on Material in Chapter 8

8.1 Which of the following are injections into the Circular Flow of Income, which are withdrawals and which neither?

(a) Saving of wages by workers.
(b) Wages paid to civil servants by the Government.
(c) Spending by foreigners on British exports.
(d) Income Tax.

(e) Investment by a firm in new machinery.

(f) A gift of £5 from a father out of his spending money to his son.

(g) Spending of wages by workers.

8.2 Cross out the inappropriate alternatives:

The Marginal Propensity to Withdraw is the *proportion/amount* of any increase in *average/total* income which is withdrawn by saving, taxation, purchase of imports etc.

8.3 If saving is the only withdrawal from the circular flow of income, and the Marginal Propensity to Save is 0.25, what will be the size of the multiplier?

8.4 Cross out the inappropriate alternatives:

National Income will be in equilibrium when the intended level of injections is *greater than/less than/equal to* the intended level of withdrawals. The equilibrium level *must always/need not* correspond to one at which all the nation's productive resources are fully employed.

8.5 Examine the following diagram:

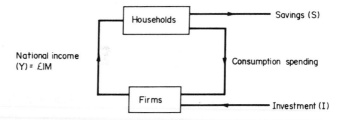

National Income is the sum of wages, rent, interest and profit. The level of National Income is in equilibrium.

The Marginal Propensity to Consume is constant at all income levels. Ignore other forms of injections and withdrawals.

(a) If the spending on consumption is £600,000 and the National Income is £1m, what will be the level of

Saving (S) . . . .

and Investment (I) . . . .

(b) Assuming that MPC is constant at all income levels and that MPC = APC (Average Propensity to Save) what size of MPC will be required to give the consumption level given in (a) above?

(c) If the MPC rises to 0.8, and assuming that MPC = APC, what will be the new levels of C, S and I which are appropriate to give a National Income of £1m?

C . . . .

S . . . .

I . . . .

(d) If I is in reality not the figure in (c) above, but is £400,000 and MPC = 0.8, to what level will National Income have to rise before equilibrium is restored? At this level what will C be?

N.I. . . . .   C . . . .

8.6 Study the following diagram; the line CC shows the level of intended consumption spending. Are the statements below *true or false?*

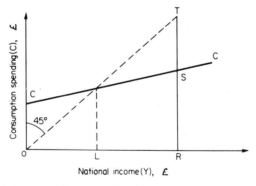

National income (Y), £

(a) Intended consumption spending increases as income increases.

. . . .

(b) At some levels of income, intended consumption spending is greater than the level of income.

. . . .

(c) Saving is positive at all income levels.   . . . .

(d) Consumption is a fixed proportion of income, i.e. the Average Propensity to Consume is constant at all income levels.

. . . .

(e) There is a diminishing Marginal Propensity to Consume.

. . . .

(f) For the level of saving to be TS the level of income would need to be OR.

. . . .

(g) If OR is the level of income, the level of consumption would be SR.

. . . .

8.7 In the diagram below state whether the following will have the effect of raising or lowering the C + I + G line and hence raising the equilibrium level of National Income:

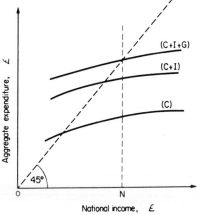

(a) increasing consumers' spending intentions by a relaxation of hire-purchase restrictions;

. . . .

(b) an increase in Government taxation without more Government spending;

. . . .

(c) additional Government spending without more taxation;

. . . .

(d) a Government programme to encourage personal saving;

. . . .

(e) raising the rate of interest on bank loans and hire-purchase agreements;

. . . .

(f) making credit from banks easier to obtain;   . . . .

(g) increase in retirement pensions and family allowances without an increase in taxation.

. . . .

8.8 Examine the diagram below: I, II and III represent three levels of aggregate demand in an economy.

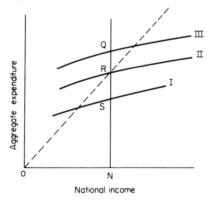

National income

If ON is the full-employment level of National Income:

(a) What is the deflationary gap by which the Government may wish to raise the aggregate demand of Curve I to achieve full employment?

(b) Conversely, what is the inflationary gap which must be closed if level of aggregate demand III is operating?

(c) If aggregate demand III is allowed to continue, what will be the consequences for the level of prices?

8.9 If there is an inflationary gap which of the following is the most promising way of closing it?

   (a) Increase Government expenditures on goods and services, leaving tax collections unchanged.

   (b) Increase taxes, leaving Government expenditures unchanged.

   (c) Increase both Government expenditures on goods and services and tax collections by the same amount.

   (d) Increase Government transfer payments from the rich to the poor.

8.10 If M = money supply

   V = velocity of circulation

   P = average price level

   T = number of transactions

which arrangement of the above letters represents the tautology which is the basis of the crude Quantity Theory of Money?

.... × .... = ... × ....

8.11 Inflation is conventionally described as having two main types of causes. What are these two types?

....

8.12 Cross out the inappropriate alternatives:

"In the Phillips Curve the *level of prices/annual percentage change in the level of prices* is shown as being *inversely related to/ proportional to* the level of unemployment, so that the higher the level of unemployment the *higher/lower* the level of inflation.

CHAPTER 9

International Trade

INTERNATIONAL trade is an extension of the phenomenon of specialisation and exchange which can be seen operating at all levels of society. No one individual in a developed economy such as the UK attempts to be wholly self-sufficient; no man will attempt to grow all his own food, make all his own shoes and clothes, build his own house, make his own car and mine his own coal. He specialises in one process, where his abilities are relatively most pronounced, using the money he earns to purchase those commodities which he does not make for himself but which are made by other specialists. This process of specialisation allows a greater quantity of goods and services to be produced, and hence a higher material standard of living to be enjoyed, than would be possible if every individual were to be completely self-sufficient.

Specialisation must always be accompanied by the exchange of goods and services between people. At the individual level in a market economy this involves earning and spending money—a very useful medium of exchange. Regions or countries can also be seen to specialise according to the abilities of their inhabitants and the natural resources of the area; hence cars are produced chiefly in the Midlands, wheat in East Anglia, sheep in the upland areas, coal in the north of England and Wales etc. Evidence of regional trade resulting from specialisation is obvious; simply look at any motorway and see goods being shifted from one area to another.

International trade can be considered as an extension of regional trade. Countries forsake self-sufficiency because specialisation and trade result in higher material standards of living for their inhabitants. In market-based economies international trade will occur simply through the natural inclination of businessmen and individuals to buy from the cheapest source. If imported Israeli oranges were cheaper in England than oranges grown in glass-houses in Britain a trade in oranges would soon be set up. On the other hand, British cars might be cheaper in Israel than Israeli-built vehicles, so trade in cars might also occur. Trade between Britain and Israel in oranges and cars would benefit both countries.

Trade involving the movement of goods can be seen as an alternative to the movement of factors of production in pursuit of their optimal allocation. If the marginal product of a factor is greater in one area than another it will be beneficial in terms of maximising output to switch units of the factor from the low productivity area to the high productivity one until the marginal products are the same in each. This is one more example of the use of the Principle of Equimarginal Returns. The price system will encourage movement in this direction because factor earnings will be higher in the higher productivity area.

In practice the geographical movement of factors is limited. Despite publicity given to the southward drift of labour in England to find jobs or better wages and the "brain-drain" of scientists and technicians from the United Kingdom to North America, most labour is reluctant to move far and frequently within a country, and is even less willing to take the major step of moving abroad, with its inherent unfamiliarity and possible currency and language problems. Most physical capital is immobile in the short run, although the reallocation of capital over a number of years between countries is within the powers of large multi-national firms. Added to this, governments may impose restrictions both on the international movement of labour and of capital, though between Member States of the European Community such barriers have largely been dismantled. Land surface and climate, both particularly important in agricultural production, are obviously geographically immobile. While the geographical movement of naturally occurring substances such as oil or iron ore, where a country sells part of its stock of productive resources in exchange for immediate purchasing power, is an important element in international trade, to a large extent countries have to put up with the resources with which they find themselves. The problem then becomes one of reallocating their productive abilities in such a way that each country specialises in producing those goods in which it has a comparative advantage and cuts back its production of the others. Trade in goods and services accompanies this process of specialisation, and the ease with which factors of production can be switched between industries—occupational mobility—becomes an important determinant of the advantages flowing from trade.

Diversity of Resources Between Countries

No two countries are identical in the productive resources they possess. It follows that the ease with which they can produce the various items which their people want also varies; in other words their production possibilities differ. It is conceivable that a country *could* produce internally some of almost every commodity that its inhabitants demanded—grapes under glass in Scotland, petrol from coal in Germany etc.—but such self-sufficiency denies the great advantages that can be earned through specialisation and trade.

Countries differ in the *natural resources* they contain. For example, coal is mined in the USA, UK, Germany etc., while other countries may have none. Large iron ore deposits occur in USA, France, Sweden and Russia; diamonds in South Africa, copper in Zambia. Climates differ, so that citrus fruits can be readily grown in Mediterranean countries, coffee in Brazil and rubber in Malaysia. It is obviously beneficial for Britain to trade with those countries which possess natural resources or climates which enable them to produce things which Britain cannot or can only with extreme difficulty; hence we sell cars to Zambia in exchange for copper, armaments to Israel in exchange for oranges, manufactured goods to the West Indies in exchange for tourism.

Countries may have acquired certain *skills* which act as the basis for trade—the Swiss watch trade or wine-making skills of France are examples—or for historical reasons have built up stocks of *capital*, such as the shipyards of Scotland or the concentration of banking and insurance institutions to be found in the City of London.

Theory of Comparative Advantage

What is less obvious is that trade between countries which are basically similar in the resources they embrace can be beneficial. For example, trade between the various members of the EEC has raised living standards, yet the member countries are similar (though by no means identical)—Belgium and Holland for example are much more alike than England and Cyprus, yet trade is beneficial to both.

The benefits from specialisation and trade can be illustrated by taking a simple model of a world which consists only of two countries (A and B), each of which has its resources fully employed in producing two commodities, bicycles and oranges, and examining the *Opportunity Costs* of increasing production of each of these goods in the two countries. The opportunity cost of producing one extra case of oranges in terms of the *number of bicycles* foregone is as follows:

> In Country A: 0.2 bicycles
> In Country B: 0.5 bicycles

Country A is said to have a *Comparative Advantage* in the production of oranges, since its opportunity cost of expanding its production of oranges is the lower.

Alternatively we can examine the output of oranges which must be lost if bicycle production is expanded in the two countries. *The number of cases of oranges* foregone if one extra bicycle is produced is:

> In Country A: 5 cases of oranges*
> In Country B: 2 cases of oranges

Country B has a *Comparative Advantage* in the production of bicycles since expansion in that product involves a lower opportunity cost in terms of oranges foregone.

Note that one country cannot have a comparative advantage in *both* commodities.

Specialisation

Specialisation in the goods in which a country has a comparative advantage will permit a greater total quantity of both goods to be produced. Let us assume that *before trade* countries A and B produce and consume the following:

	Cases of oranges	Bicycles
Country A	500	300
Country B	800	700
A and B together	1,300	1,000

This Table (or schedule) shows both the quantities involved in production and consumption since, without trade, the inhabitants can only consume what they produce.

---

* These figures are the reciprocals of those above.

Let us assume that A expands its production of oranges, the good in which it has a comparative advantage, by 30 cases. By doing so it will have to forego the production of 6 bicycles (remember the opportunity cost of production was 0.2 bicyles per 1 case of oranges). Similarly, let B expand its production of bicycles, the good in which it has a comparative advantage, by 10 cycles, reducing its output of oranges by 20 cases to do so. The new production schedule will be as follows:

	Cases of oranges	Bicycles
Country A produces	530	294
Country B produces	780	710
A and B together	1,310	1,004

Note that specialisation has resulted in the combined output of the two countries expanding so that 10 extra cases of oranges are produced (1,310 as opposed to 1,300) and 4 extra bicycles (1,004 as opposed to 1,000).

Trade in Surpluses

When countries specialise the balance of production between the two goods will be wrong for domestic consumption—in Country A too many oranges will be produced and too few bicycles. However, this balance can be restored by the export of excess oranges in exchange for bicycles produced in Country B, which will find itself initially with a deficiency of oranges and a surplus of bicycles. Obviously some exchange ratio between oranges and cycles will have to be arranged between the countries. It will be found that this international exchange rate will settle by bargaining at a level somewhere between the internal opportunity cost ratios, i.e. somewhere between 5 cases of oranges per bicycle (A's opportunity cost of production) and 2 cases per bicycle (B's opportunity cost of production). Let us assume that the international exchange rate settles at 3 cases of oranges for 1 bicycle.

At this ratio Country B will be able to trade 7 out of the 10 extra bicycles it produces after specialisation in exchange for 21 cases of oranges. Country A will, reciprocally, trade 21 of its extra 30 cases for 7 bicycles. After trading has occurred the quantities of the goods which each will be able to consume (i.e. the consumption schedules) are as follows:

	Cases of oranges	Bicycles
Country A consumes	509	301
Country B consumes	801	703
A and B combined	1,310	1,004

Comparing this schedule with the situation before specialisation and trade shows that both countries now consume greater quantities of both commodities. Both countries have benefitted. Combined production is increased. When countries are not forced to be self-sufficient i.e. no longer compelled to produce themselves what they wish to consume, they can specialise in those goods in which they have a comparative advantage, trading part of their excess production for goods which are the speciality of other countries.

Comparative and Absolute Advantage

Note that no reference has been made to the level of productivity or efficiency of either country. One could have been a place where output per man, and hence wages, were high and the other a low-output country. Output per man (or per some other unit of resource) indicates which country has an *absolute* advantage—Country A may produce both more oranges and more bicycles per man than Country B—but this is of no relevance to the advantages to be gained from trade. What determines the advantages from trade is comparative advantage—which means differences in the opportunity costs of production.

The irrelevance of absolute advantage can be illustrated by an example of specialisation at the individual level. A solicitor employs a secretary to type his letters. If he discovers that, as well as being a much better lawyer than his secretary could ever be, he is also a marginally better typist, will it pay him to sack his typist and spend half his day typing his own letters, cutting back his legal practice to allow him time? The answer is no: he should keep his secretary so that he can spend all his time doing the job in which he has a comparative advantage—law. By doing so he could earn more, even though he has to pay a secretary. The solicitor has an *absolute* advantage both in law and in typing, but this is irrelevant to specialisation. He has a *comparative* advantage in law, or, alternatively, a comparative disadvantage in typing—an hour of typing has a higher opportunity cost in terms of earnings foregone for the lawyer than for the secretary—and it is comparative advantage which leads to benefits through specialisation and exchange at personal, regional or national levels.

Countries do not carry specialisation to extreme for a variety of economic and political reasons. To illustrate the fundamental economic reason we must return to our simple model.

Limits of Specialisation

Country A cannot expand its production of oranges indefinitely, nor B its production of bicycles. While initial expansion in the direction indicated by comparative advantage may be easy, further specialisation will be increasingly difficult. Country A will find that, to produce successive additional cases of oranges, it has to give up producing an increasing number of bicycles i.e. the opportunity cost of orange production will rise but the opportunity cost of cycle production will fall. In Country B expanding cycle production will cause its opportunity cost in terms of oranges foregone to rise, but the opportunity cost of orange production will fall. Expansion of orange production may bring with it diseconomies of scale or the land, labour and other factors switched to orange production may tend to be less suitable for orange production than they were for cycle production. The quality of factors is not uniform, and the grades giving the greatest benefit in terms of oranges gained per cycle foregone will tend to be transferred first, the less suitable later. This can also be illustrated at the farm level. If a farmer wishes to increase his acreage of wheat at the expense of another crop, potatoes, he will tend to select the most suitable of his land first; this is not necessarily the land which gives the highest wheat yield but is that which gives him most of the expanding crop per unit of the crop which is being contracted, i.e. the largest ratio between tons of wheat gained to tons of potatoes lost. Continuing wheat expansion will involve transferring land which gives a progressively diminishing benefit in terms of wheat yield for every ton of potatoes foregone.

You will recall that the reason why specialisation and trade occurred was that opportunity costs of production differed between countries. However, the very process of specialisation causes opportunity costs to change and converge, and a point will be reached at which the opportunity costs are the same in both countries. Beyond this point further specialisation will not occur.

Geometrical Representation

A geometrical representation may be helpful. The accompanying Fig. 9.1 shows the production possibility curves for bicycles and oranges for the two countries, A and B (curves PP).

These convex curves imply that the marginal rate of substitution of products (i.e. the slope of the curve) is not constant; increasing the production of either good by one unit requires progressively larger quantities of the other to be sacrificed.*

Before trade each country is at point A on its PP curve. At these points the MRS of products (i.e. the opportunity costs indicated by the slope of the PP curves at point A) differ. With the opening of trade each country adjusts its balance of production until it arrives at point B, at which the MRS of products (or opportunity costs of production) are the same; this MRS will also correspond to the rate of exchange of bicycles for oranges between the two countries. By the process of trade, countries can exchange their surplus production and consume the combination C; for each country point C lies further out from the origin than A, showing that specialisation and exchange has enabled them to enjoy both more oranges and more bicycles than they could before trade.**

---

* For a further illustration of this point see the Product-Product relationship described in the chapter on Production Economics (Chapter 5).

** Indifference curves relating to the whole populations of A and B are being implied here although not shown. Also, because the geometrical approach was required to fit the earlier numerical model, the possibility that after specialisation and exchange a country need not consume more of *both* commodities to reap the greatest rewards from trade has not been explored. For these points see Ritson (1977).

FIG. 9.1   *A Two-Country, Two-Commodity Model of the Gains
from Specialisation and Exchange*

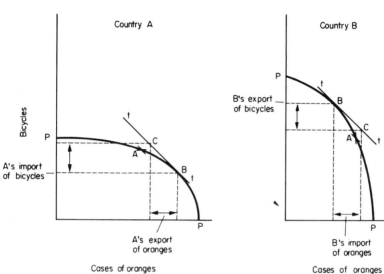

Extension of the Simple Model

So far we have dealt with only two countries and two commodities and have shown the benefits which can flow from specialisation and trade. In the real-world, however, many countries and many goods are involved.

(i) Many countries:

Trade between pairs of countries is termed *bilateral* trade (such as is met in bilateral trade agreements whereby two countries agree to buy a certain amount from each other) and between many states is called *multilateral* trade. Unhindered multilateral trade is more beneficial than bilateral trade. This is easily illustrated by Fig. 9.2 below where trade flows between three countries.

If trade could only go on between pairs of countries perhaps with the countries involved in this example no trade at all would occur. The UK might not want anything produced by Chile in exchange for the mining equipment. Similarly New Zealand might be unable to sell anything to

FIG. 9.2   *Multilateral Trade*

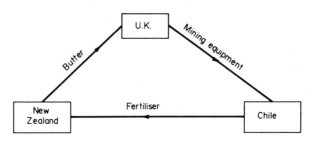

Chile in exchange for fertilisers. By allowing multilateral trade, in which the value of trade between any two countries need not balance but combined imports and combined exports for any one country will balance, the benefits of specialisation and exchange can be more fully realised.

(ii) Many commodities:

Within a country it should be possible to estimate the opportunity costs of the whole range of its products in terms of a "reference" product. These opportunity costs can then be compared with a similar set from another country. For example, Fig. 9.3 shows the opportunity costs of a range of goods in two countries in terms of food production. In Country A far less food has to be sacrificed in order to produce one extra unit of clothes than in Country B. By comparing the opportunity costs of each good in terms of food foregone in the two countries, it is possible to "rank" comparative advantages; in the Table, Country A's comparative advantage is most pronounced in clothes production but least pronounced in cigarettes. Alternatively one could conclude that Country B's comparative advantage is greatest in cigarettes and least in clothes. Where this "rank" is split for specialisation will depend on the total demand and supply situation. Note that it is not absolute advantage which determines the ranking, but comparative advantage.

Transactions Involving Currencies

So far in this section, trade has been treated as if it were conducted on behalf of nations by their governments. But this is, of course, not what happens. International trade, at least between Western market or

FIG. 9.3   *Ranking of Comparative Advantage*

Increasing production by 1 unit causes the production of food to be cut back by	Country A units	Country B	Ratio of Opportunity Costs (B/A)	Ranking of Opportunity Costs
Wine	10	20	2.0	3
Cigarettes	6	2	0.3	5
Beer	1	3	3.0	2
Clothes	2	10	5.0	1
Food	1	1	1.0	4

mixed economies, is basically undertaken by individuals and firms who find that they can buy and sell goods across national boundaries with financial advantage. The reason why a businessman in England buys a Dutch-made lorry is probably because the foreign vehicle is cheaper or better or more available than one made in Britain. Americans buy Scotch for similar reasons. Although the role of Government in trade is essentially a secondary one—encouraging trade or restricting it for economic, political or social purposes—it also enters directly into trade by purchasing Government supplies (such as armaments) from abroad, maintaining troops in foreign stations etc.

Trade across national boundaries is complicated by the fact that each country generally has its own currency. If a British manufacturer sells a machine to a Japanese businessman, the Briton will not welcome a payment in Yen, because workers in his British factory will not accept their wages in Yen. There must be some mechanism whereby Yen can be exchanged for £ Sterling. In other words, for trade to occur there must also be the facility of international currency exchange; multilateral trade will require the facility to exchange units of one currency for any other currency. Currency exchange is undertaken in the Foreign Exchange Market, in which commercial banks and central banks (like the Bank of England) are active. It is a market because, for each currency, both a demand and a supply exist; the interaction produces a price for each currency which can be expressed with reference to any other of the currencies in the market. Hence we can talk of the exchange rate between Sterling and Dollars or between Sterling and Yen or between

Dollars and Yen; each of these three exchange rates will be closely linked to the other two.

For the moment we will concentrate attention on the way the market for currencies is related to trade, although later the effects of international capital movements and the interventions by central banks will have to be considered too. Take the example of the exchange rate between £ and $; the demand for £ (and also the supply of $) arises from American businessmen going to their local banks in America and arranging to buy £ Sterling in exchange for $ so that they can purchase goods from British firms. The supply of £ (and the demand for $) arises from British businessmen wishing to sell £ in exchange for $ so that they can import American goods. There will be some rate of exchange between £ and $ where the supply and demand of each is in equilibrium. If this exchange rate is allowed to alter, and is not held rigid by Government order for political or social reasons, its flexibility will ensure that the trade in goods and services normally flows in the direction suggested by comparative advantage. A simple model will show this.

Flexible Exchange Rates

Take the example of a world consisting of only two countries (the UK and the USA) and with only two commodities (food and clothes). It can be shown that, under perfect competition, the relative prices of commodities are indicative of the quantities of resources used in their production. Fig. 9.4 below shows the prices of home-produced food and clothes in the two countries before trade starts up.

In the USA a unit of clothes is 3.3 times the price of a unit of food, indicating that in the USA producing a unit of clothes takes 3.3 times the quantity of resources required to produce a unit of food. In the UK it only takes 1.25 times as much; the UK therefore has a comparative advantage in clothes production since producing one extra unit of clothes involves sacrificing a smaller quantity of food in the UK than in the USA.

If on the international currency exchange market the rate between the two currencies is initially set at $10 = £1, the Americans will be able to undercut the UK home prices of both food and clothes. Ignoring transport costs, the cost in the UK of an American good can be easily calculated using the exchange rate to be as follows:

Price of a unit of USA food in the UK   = 10p (in UK currency)
Price of a unit of USA clothes in the UK = 33p (in UK currency)
UK citizens will naturally try to buy the cheaper imported goods. To buy them they must obtain dollars to exchange for the food and clothes. This demand for dollars will bid up the price of dollars in the currency exchange market, i.e. the exchange rate will alter, with the dollar appreciating. Eventually the value of the dollar will rise so that US goods only undercut the home market in one commodity. If this rate is $2 = £1, the cost of American goods in the UK will be as follows:

Price of a unit of USA food in the UK   =   50p (in UK currency)
Price of a unit of USA clothes in the UK = £1.65p (in UK currency)

Fig. 9.4   *Prices of Domestically Produced Goods before Trade*

	USA	UK
Food	$1 per unit	£1 per unit
Clothes	$3.3 per unit	£1.25 per unit

At this level of prices UK citizens will find imported food cheaper than home produced food, but foreign clothes more expensive. Similarly, USA citizens will find imported clothes cheaper and imported food dearer than that produced at home. This situation will not only produce a two-way flow of goods but also, as a by-product, there will be both a supply and a demand for each currency. In this model the price and exchange rate mechanisms have ensured that each country specialises in the commodity in which it has a comparative advantage. The absolute level of prices before trade was irrelevant.

If UK citizens suddenly take an increased liking to American goods and try to buy more, the demand for $ will increase (and the supply of £ will also increase) so that the value of the $ will be bid up, i.e. the exchange rate will alter, with the $ appreciating (or, which is the same, the £ will depreciate). This will make American goods in Britain more expensive and reduce the demand for them, tending to restore the balance between the total value of imports and exports.

Domestic inflation will also be reflected in changes in the exchange rate. As long as the exchange rate is allowed to move freely, inflation will not affect the directions in which traded goods flow. This is illustrated simply: assume the level of all UK prices goes up ten times but the *ratios* of prices in each country remain unaltered. The new level of prices will be

	USA	UK
Food	$1 per unit	£10 per unit
Clothes	$3.3 per unit	$12.5 per unit

At $2 = £1, which was the exchange rate at which trade had been occurring before UK inflation, UK citizens find both USA food and clothes cheaper than the home produced items. The demand for dollars rises and the exchange rate is bid up to a level where only USA food is cheaper than UK food, USA clothes becoming more expensive than UK clothes, and trade is re-established. The new rate will value the dollar ten times higher than previously, i.e. $2 = £10. (This can also be thought of as a fall in the exchange value of the £, to a tenth of its pre-inflation level.)

This model again emphasises the point that the absolute price levels are irrelevant to specialisation and trade, which depend on differences in comparative costs.

Assumptions in the Theory of Comparative Advantage

The theory of comparative advantage shows that higher material living standards can be attained through trade. However, the theory makes a number of assumptions which are not entirely borne out by the real-world, and acknowledging these assumptions helps to explain why, despite the apparent advantages to be gained from unhindered trade, most countries use devices such as tariffs or quotas to restrict it. (A tariff is a tax on imported goods which, in effect, raises their price to the importer, and a quota is a restriction on the quantity of imports permitted.)

1. *Static or dynamic conditions:* we have assumed that a country with

a comparative advantage will always possess it. But in reality things are dynamic; for example, less developed countries may not yet have a comparative advantage in, say, steel manufacturing, but if such industries could get established they might develop a real comparative advantage. Protection of such "infant" industries may well be justified in the long run as, when young, they will usually be undercut by the lower prices of the established industries in other countries. However, countries must beware of protecting infants which never develop and require permanent protection.

Technical advances also tend to shift comparative advantage. In the 19th century Britain had a comparative advantage in textile production, but the development of artificial fibres and new machinery, together with changes in other sectors of the economy, have apparently shifted the comparative advantage to countries such as Taiwan. The factors of production embodied in the British textile industry therefore need to be transferred to other industries as competition undercuts the prices of British-made articles.

2. *Full employment of resources:* specialisation and exchange involves the contraction of those industries in which the country *has not* a comparative advantage. This means that factors of production—labour, capital, land and business ability—have to be switched from declining to expanding industries.

Much depends on the rapidity with which such switches of factors are made. A sudden decline in, say, the ship-building industry will simply swell the ranks of unemployed shipyard workers. The potential advantages from trade will be largely lost unless this labour is taught new skills and is absorbed by the expanding industries. Indeed, without this reabsorption it is possible that the nation may be worse off, particularly if unrest caused by redundancies leads to strikes. The loss to national income by such action can easily be greater than the potential gains from trade. Similarly, capital goods in industries shrinking rapidly under pressure from overseas competitors frequently becomes unemployed because of their specific nature (e.g. textile machinery) and do nothing to aid the expansion of the industries in which the UK possesses a comparative advantage.

The moral is that factors need time to adjust—labour to be retrained, capital to wear out and be replaced by other types, land to be switched to other uses—and a too-rapid change caused by opening up trade may cause factor unemployment which more than offsets any potential gains from freer trade. To prevent this waste and social distress, short-run restrictions on trade may be justified to give factors time to be switched.

Distribution of the gains from trade: while trade can be shown to be beneficial to nations as a whole, given certain assumptions, not everyone benefits to the same degree. The owner of a British business which could suffer from competition from abroad may take a poor view of Government policies to promote international trade—his prices will be undercut and he will need to look for other opportunities of using his ability or resources. He will advocate more tariffs, quotas etc., to protect his own interests. He will support a Buy British campaign and attempt to make buyers of foreign produced goods feel unpatriotic. He will attempt to extract subsidies from the Government by persuading them that an industry vital to the national interest is being threatened.

The specialisation and exchange which trade involves has been said to benefit many people a little but hurt a few people a lot. The few sufferers will readily band themselves into pressure groups to protect their own interests but it is far more difficult to organise the rest of the population, each of whom benefits only marginally, into an effective lobby to press for more trade. However, if at the whole-economy level, trade is beneficial, it must be at least theoretically possible to adequately compensate the sufferers from the gains of the beneficiaries, with something remaining left over as a net gain.

Similarly, there is no guarantee that the broad factor groups will receive benefits in the same proportion; for example, the owners of capital may find themselves much better off if comparative advantage indicates that a country should expand its production of goods which require much capital. On the other hand the rewards to labour may not increase as much. Specialisation and trade generally work so that the factors which were initially the most scarce become less scarce as specialisation proceeds, so that the share of the nation's income they receive will tend to fall. If this produces a distribution of income which society considers unjust, incomes can be redistributed by differential taxation and benefit payments.

Arguments in Support of Trade Restrictions

To Protect against Competition from Cheap Labour Countries:

The arguments put forward for protection often place emphasis on the harm the economy can suffer through one industry being undercut by foreign competition from countries where labour is cheap. Such competition is said to be "unfair". This argument is not valid. It is true that specialisation and trade may reduce the *relative* earnings of labour in "dear labour" countries, i.e. its share of national income may decline and the absolute level of income of some individuals may fall. However, as the levels of national income in both the "cheap" and the "dear" labour countries will increase through trade (after re-allocation and adjustment), it is quite possible, although not necessarily done, to arrange via the tax and benefits system in each country that everyone benefits.

To Prevent Dumping:

Tariffs are often advocated to prevent dumping, that is the export and sale at very low prices of a surplus which a country might have accumulated as the result of some domestic policy. For example, before Britain joined the European Economic Community there were occasions when French "Cheddar-type" cheese was sold to British shopkeepers for retailing at prices considerably below that of British-produced cheese, disrupting the demand for British cheese and causing difficulties for the dairy industry. This dumping was only temporary, yet its possible effect on the British dairy industry could have been longer term—if serious disruption and bankruptcies had occurred, the housewife might have found herself eventually paying more for British cheese once French dumping had ceased. Tariffs or quotas are tools which can be used against such disruptive dumping.

On the other hand, if Eastern European states are willing to sell their motor cars in Britain at prices which are less than the costs of production in order to earn pounds Sterling, and they are prepared to do this on a long-term basis, is it not wise to take advantage of their generosity?

To Raise Revenue:

The amount of revenue which a tariff raises depends on the price elasticity of demand of the good on which the tariff is imposed. Increasing the tariff on goods with high price elasticities of demand could result in a *fall* in tax revenue as the rise in price will severely choke off the demand for imports. In performing their revenue-raising role, tariffs protect home industries from competition from cheaper foreign goods and hence prevent, *ceteris paribus*, the optimum international allocation of resources. To prevent such a distorting effect, some would suggest that a tax equal to the tariff should be placed on home-produced goods. Taxes on goods where no significant UK production is possible, such as tobacco and brandy, have long been used as a source of State revenue. With these commodities no distortion between domestic and foreign production of each particular good arises since domestic output will always be constrained by climatic factors to zero, or to very low levels.

For Retaliation:

An understandable emotional reaction to, say, the USA erecting a tariff wall against UK goods is for the UK to place similar tariffs on USA goods coming to Britain. Yet fundamentally this is about as sensible as digging holes in British roads to damage the cars of Belgian holiday-makers visiting Britain because British cars are sometimes damaged by poor Belgian roads. The *threat* of a retaliatory tariff may, however, sometimes be sufficient a deterrent to prevent a country contemplating such a restriction from implementing their action.

To "Export" Unemployment:

We have agreed that a temporary restriction on trade may be necessary to prevent the unemployment of factors in the short run, where this is caused by their inability to shift from declining to expanding industries. However, where the whole country is in a slump, with widespread unemployment caused by a general deficiency in demand, countries have occasionally been tempted to help home employment by imposing import quotas or by erecting a tariff fence, thereby cutting down imports and hopefully expanding home production and home employment to take advantage of the drop in imports. But this ultimately causes unemployment abroad and excites retaliatory action by other countries. The "export" of unemployment was practised during the inter-war period, and resulted in tariff walls being erected and a rundown in trade. In order to discourage such a situation from arising again, after the Second World War a large number of countries (including the UK and USA) signed the General Agreement on Tariffs and Trade (GATT). Under GATT rounds of negotiation continue to be held, with the aim of dismantling existing tariffs and other trade barriers and of preventing additional distortions.

For Social or Defence Reasons:

Britain has found that, during periods of "conventional" warfare, a home food supply is a valuable asset. For this strategic defence reason, measures to protect British agriculture against international competition in peacetime may be justified. On a more general level, agricultural protection might be used to maintain a prosperous rural population with many small farms, since a depopulated countryside and large-scale farming practices may be neither politically nor socially acceptable, although it is debatable whether trade restrictions are the cheapest way of achieving the desired results. Within the European Community, where unrestricted trade in agricultural products was intended to apply, regional specialisation has not been allowed free reign. For example, farming in hill areas would largely disappear if subject to undiluted market forces. For social (and political) reasons these producers are protected, using special payments to farmers in Less Favoured Areas (LFAs). Also for social reasons trade restrictions are placed on the imports of drugs—in the case of tobacco the underlying reasons seem to have moved towards that of improving the standard of the nation's health from one purely of raising revenue. In the past, member nations of the "family" comprising the Commonwealth have enjoyed trade preferences (i.e. lower tariffs etc., than non-members) for a mixture of social, political and economic reasons. Now this applies to Member States of the European Community.

Summarising restrictions on trade, it can be concluded that many of the arguments put forward for "protectionism" cannot bear close scrutiny. While short-term trade restrictions can be justified where waste would be caused by attempts at sudden adjustments in the economy, long-term restrictions are likely to be harmful unless any economic benefits to be gained are more than offset by social (or defence) deteriorations. It is necessary to examine closely the real motives behind any trade restriction.

Balance of Payments

Deficits or surpluses in the Balance of Payments form the material for banner headlines in newspapers and strong political debate. The preservation of an acceptable Balance is an important objective of national economic policy, and economic growth and low unemployment have frequently been sacrificed in its interest. Before discussing the reason for its prominence we must briefly describe the elements of the Balance. The Balance of Payments is usually divided into four categories; the trade in "visibles", the trade in "invisibles", flows of capital and monetary movements. *Visible trade* is comprised of items which literally can be seen, e.g. cars, television sets, food etc. Our imports increase the demand for foreign currency (because Britons need the currency to pay the foreign exporters) and increase the supply of Sterling. Our exports increase the demand for Sterling (because foreigners need Sterling to pay for British goods) and increase the supply of foreign currency. The balance of the value of visible imports and exports is called the *Balance of Trade*. It is negative if the value of visible imports exceeds the value of visible exports. If this negative figure becomes greater (i.e. grows from − £10m to − £20m) the equilibrium between supply and demand for currencies will tend to alter, with the exchange value of the £ falling.

The *Terms of Trade* is not to be confused with the Balance of Trade. The *Terms of Trade* compares relative movements in indices of the *prices* of exports and imports.

$$\text{Terms of Trade} = \frac{\text{Index of export prices}}{\text{Index of import prices}}$$

The Terms of Trade are said to "improve" if the ratio increases, i.e. if the price per unit of our exports increases or if imports become cheaper. On the other hand, the Terms deteriorate if import prices increase with no equivalent rise in export prices, as happened when the price of oil was suddenly increased in 1973/4 by the oil exporting countries. Following the deterioration a greater volume of British goods had to be exported to pay for each barrel of imported oil than before. *Invisible trade* comprises the following: banking and insurance services; tourism and travel expenditure; Government spending abroad to maintain troops on foreign stations and embassies etc.; shipping and civil aviation services; income from investments abroad and payment to foreigners arising from investments they hold here; private gifts such as occur when foreign workers send money back home to relations, and other miscellaneous payments for the hire of foreign films etc. While this group of "invisibles" is rather a hotch-potch, they have the following characteristics in common:

(1) no tangible goods are *directly* involved (even with the hire of USA films, what is being bought is the right to see the film, not the ownership of the film itself);

(2) the items are essentially for current consumption as opposed to investment;

(3) spending by Britons increases the demand for foreign currency and hence tends to lower the value of Sterling (or increase value of foreign currency, which amounts to the same) while spending by foreigners, say, by visiting Britain, tends to increase the value of Sterling.

Taken together, the Balance of Trade plus the balance in invisible earnings comprise the Balance of Payments on Current Account. This is illustrated in Fig. 9.5. It is this Current Account Balance of Payments which is the subject of much political controversy. A deficit, with the value of imports exceeding exports, will tend to cause the value of the £ to fall, although this may be counteracted by what is happening to flows in the Capital Account and by monetary movements.

Capital flows: some spending on foreign goods is not for current consumption, but is for investment. For example, if a Briton buys a farm in France he is hopefully making an investment in an asset which will yield him an income. (Incidentally, this income will subsequently appear as an invisible earning in the Current Balance of Payments.) Similarly a Briton may buy shares in a foreign firm. The purchase of foreign assets will increase the demand for foreign currency and increase the supply of Sterling—i.e. it will have a depreciating effect on the value of the Pound similar to that of the purchase of imports. Investment by foreigners in British assets will of course have the reverse effect. In a time when the Balance of Payments on Current Account is negative, an outflow of capital will tend to aggravate the downward pressure on the value of Sterling and so investment abroad may be restricted by the Government. It must not be forgotten, however, that this means the *income* from such investments in future years is being foregone. The combined balance on Current and Capital Accounts produces the Total Balance of Payments.

Monetary Movements

In a situation where the value of Sterling in terms of other currencies was determined solely by the demand and supply for currencies arising from trade, if Britons tried to import a greater value of goods than they exported the exchange rate of the Pound would fall, making imports more expensive, choking off the demand for them, and making our exports more attractive in foreign markets. Thus the imbalance would tend to be corrected. Because trade is not an instantaneous process, and because of complications involving the demand elasticities of our exports and imports, the depreciation in the exchange rate for Sterling, while bringing the demand and supply for the Pound into balance, would not completely correct a Balance of Payments deficit, at least not in the short run. Nevertheless, over a few years such a depreciation could be reasonably expected to contribute to a fall in the deficit. However, it was conventional UK Government policy until 1972 to maintain the international exchange value of Sterling at or very close to a publicly specified figure since this was reputed to impart confidence to traders; international trade, it was argued, depends on minor cost advantages which could easily be wiped out by variations in exchange rates. This was called a "fixed rate" policy.

If the Government has decided to allow only small changes in the exchange rate of Sterling it must be prepared to take action if adverse movement in trade tend to depreciate the Pound. This is done by the nation's central bank, the Bank of England, holding in its Exchange Equalisation Account a reserve of foreign currencies and gold. When the value of the Pound falls the Bank increases the demand for Sterling by selling foreign currency or gold and buying Pounds. Conversely

FIG. 9.5 *Balance of Payments*

---

1. *Current Account*

Debits (−)	*Credits* (+)	*Balance* (+ or −)
Visible imports (cars, TV etc.)	Visible exports (clothes, tractors etc.)	Balance of Trade (£a)
Invisible imports (Holidays abroad, interest to foreign investors, govt. spending abroad etc.)	Invisible exports (Banking and insurance, foreign tourists, sea and air transport etc.)	Balance of invisible earnings (£b)
		Balance of Payments on Current Account (£x = £a + £b)

2. *Capital Account*

Capital flow out (−)	*Capital flow in* (+)	
Official and private investment abroad	Official and private investment by foreigners in UK	Net capital flow (£y)

Combined Accounts	*Total Balance of Payments*
Current Account balance + Capital Account balance	(£x + £y)

Monetary movements

to cover B of P deficit or surplus

(a) increase (−) or decrease (+) in UK reserves of gold and foreign currencies

(b) increase (+) or decrease (−) in borrowing from International Monetary Fund

(c) increase (+) or decrease (−) in borrowing from foreign banks etc.

Combined monetary movements	(£x + £y)

---

when the value of the Pound has risen to a certain level above which the Government thinks it should not rise further, the Bank sells pounds and takes in foreign currency to increase its reserves. If the Bank's reserves of gold and foreign currency dwindle to a very low level, through it having to buy Pounds more often than selling them, then the Bank has to borrow from banks in foreign countries or from the International Monetary Fund (IMF), an organisation set up after World War II to facilitate trade; one important way it does this is by acting rather like a super-national bank in granting credit to member countries with temporary Balance of Payments difficulties. If the Total Balance of Payments is, say, £500m in deficit then this will have to be covered by a £500m decrease in gold and foreign currency reserves, *or* a £500m loan from the IMF or from foreign banks. This compensating change in the stock of currencies comes under the heading of Monetary Movements.

If the Pound is allowed to "float" i.e. its value to be determined by the supply and demand for Sterling, then the Account will only intervene in the exchange market on those rare occasions when the Pound's value is *temporarily* affected by such things as a General Election. This policy could be called a "Managed Float". If the Government permits the market value to fluctuate, but only within narrow limits, then the Exchange Equalisation Account may find itself buying or selling with much greater frequency. Furthermore, if the band of managed exchange rates overestimates the value of Sterling, the Account will soon find itself very short of reserves of gold and foreign currency. It is then necessary to let the Pound depreciate to an exchange rate which more truthfully reflects its market value.

Management of exchange rates in the past was made more difficult for Sterling because it was traditionally a "reserve" currency. A number of other countries collectively described as the Sterling Area, and including many members of the Commonwealth, tended to keep their national currency reserves in Sterling as an alternative to gold because much of their trade was with Britain and because of Sterling's traditional stability. These Sterling Balances were largely invested as short-term loans to the British Government. Although Sterling's reserve role has virtually ended, due in part to entry to the EEC, London's position as a major world banking centre still attracts funds from abroad. Any rumour of a sudden depreciation of Sterling, an upward movement of another currency such as the Yen, or the existence of substantially higher interest rates elsewhere may cause these holders of Sterling investments to sell them and buy other currencies or gold. Any depletion of the UK's gold and foreign currency reserves because of a balance of payments deficit will of course be aggravated by such a "flight from Sterling".

In order to achieve stability in exchange rates many EEC Member States have linked their currencies in the European Monetary System. If exchange rates start to move outside narrow agreed bands, collective action by national banks is used to combat these pressures—the burden of support buying and selling is shared. Provision is made for periodic changes in the agreed bands, termed "realignments". By 1989 the UK had not joined this part of the EMS, one reason stated by the Government being the reduction in the UK's freedom to manage its own economy to meet domestic problems such as inflation.

Confidence is clearly an important issue in trade and monetary movements. Loans to Britain by foreigners to cover Balance of Payments deficits will only be made if they have confidence that the deficit is temporary and that internal policy, such as inflation control, and external policy, such as currency exchange rates, are so set that a long-term balance of payments is feasible. Investment from abroad will again be encouraged by a stable domestic and trade position. Confidence will be seriously undermined if the exchange rate of a currency is persistently under downward pressure through a fundamental imbalance in trade, and this loss of confidence may do more harm than the original imbalance. Governments, then, are intensely interested in their country's trading position and have tried many methods, some effective and some largely cosmetic, to maintain international confidence in their currencies and economies.

Government Manipulation of the Trade Balance and Exchange Rates

While a Government will not feel excessively concerned over minor imbalances in trade, particularly if a floating exchange rate policy is adopted, the UK Government still considers it part of its function to engage counter policies if the fundamental trade trend is adverse and downward pressure is put on the exchange of Sterling. This it does by measures such as stimulating exports and encouraging foreign visitors and other invisible earnings. In the short run it can also restrict imports by quotas, tariffs etc. (although these measures are, theoretically at least, now largely impossible for the UK under EEC regulations), restrict domestic Hire Purchase credit so that less imports are purchased and more goods are available for export and restrict the demand for foreign currency, for example, by placing a limit on how much cash tourists going abroad can take. A domestic policy of *deflation* which cuts back home demand and tends to lower the prices of British-produced goods relative to foreign goods will encourage exports and discourage imports. Import-saving industries can be encouraged to expand. All these measures are primarily concerned with the Current Account items in the Balance of Payments. It must be remembered, however, that our imports are the exports of other countries, the earnings from which they use to buy our exports; this is particularly true of Third World trading partners. Cutting down imports will eventually harm our exports, so while import restriction policy may be a temporary expedient it is probably against the national interest if carried on for too long.

Capital Account items can also be used to influence the pressure on the value of Sterling. Investment by Britons abroad can be discouraged or restricted by controlling currency exchanges for investment purposes. Investment in Britain can be encouraged by giving incentives, such as tax concessions for Japanese car firms setting up factories in the UK, making sites readily available and so on. In the 1980s interest rates had a marked effect on exchange rates. An increase in the Bank of England minimum lending rate will make the deposit of money in Britain in the form of Sterling attractive, hence increasing the demand for it. Capital and Current Accounts of course are linked—a restriction on investment abroad will eventually curtail the invisible earnings from such investments, and increasing the rate of interest to be earned by foreign monies deposited here will increase the invisible payments out on Current Account. Because all interest rates are linked, the rates for domestic borrowing will rise with an increase in Bank of England rates, resulting in domestic deflation which may not be politically acceptable if, for example, domestic unemployment is already high.

In an attempt to correct a persistently adverse Balance of Payments, the Government may allow the exchange rate of Sterling to fall. Indeed, it may have no option. At first glance this appears a cure-all, since the price of imports will rise and the price of our exports will fall. Imports should appear less attractive and exports more attractive. This is undoubtedly true, but the Balance of Payments is not a ratio of prices, but the difference between values of exports and imports—that is, price times quantity. If the price of our exports drops by 10% as the result of a currency depreciation but only a 5% increase in quantity of exports sold abroad takes place, the *value* of exports (in £) will have risen by 5%. However, if we continue to import the same volume of goods at the lower exchange rate, the total value of imports (again in £) will rise by 10 per cent and the Balance of Payments deficit will worsen.* This is almost inevitably the sort of thing which happens in the short run after a depreciation before buyers have had time to adjust to the new levels of prices. In the longer term, say two years, the deficit may show some improvement, but to be successful a fall in the exchange rate must be accompanied by elastic demand (i.e. price-sensitive demand) for exports and/or imports, preferably both. It is far from sure that the price elasticities of demand for UK exports is high enough for a downward movement in exchange rates to be of great help in correcting a Balance of Payments deficit.

Added to this problem associated with price elasticities of demand is a longer term one arising from the low income elasticity of demand which, at least until recently, the rest of the world exhibited for British exports in general. For historical reasons the UK appeared to produce an array of goods which were not in much greater demand as world income rose. This implied that the UK fell behind other Western countries with higher income elasticities for their exports in terms of growth of exports and consequently in National Income per head. A parallel can be drawn with the decline in relative incomes experienced by the agricultural sector of a developed country as National Income continues to rise (Chapter 3); this decline can be explained by the low income elasticity of demand for agricultural products.

The static demand for UK goods abroad is in contrast with the demand by UK citizens for imported goods as British incomes rise. The result is that Balance of Payments problems, caused primarily by

---

* In terms of $s, the imports to the UK will remain unchanged, but the $ value of exports will have fallen.

rapidly increasing imports, have accompanied attempts to stimulate the UK economy into growth since World War II, necessitating rapid changes of policy to correct the deficit and ending the economic expansion. The appropriate long-term action to overcome these problems stemming from unfavourable income elasticities appears to be to encourage the production and export of goods which have rapidly expanding demands on world markets as incomes grow and which UK consumers will find preferable to imported articles. The 1980s saw a move in this direction, in part the result of the happy discovery of oil in Britain's North Sea, a restructuring of industry away from traditional manufacturing and towards lighter, hi-tech industries, and a growth in the service sector, including banking, insurance and other financial services for which there is strong international demand.

Exercise on Material in Chapter 9

9.1 If a country has to give up the production of 5 tons of wheat when it wishes to expand its production of cheese by 1 ton, what is the opportunity cost of the extra cheese?

9.2 Select the correct alternatives in the following statement:

"A country is said to possess *comparative/absolute* advantage *vis-à-vis* another country in the production of a good 'X' if the expansion of production of good 'X' by 1 unit involves foregoing the *consumption/production* of a *greater/lesser* quantity of other goods in the first country than in the second."

9.3 If it can be shown conclusively that UK agriculture is the most efficient in the world, can we say that it would never pay the UK to import food?

9.4 Examine the following two-country, two-commodity model:

"Maxitry and Minitry are two islands, similar in the productive resources they contain. The standards of living in Maxitry are higher than in Minitry because the inhabitants of Maxitry are better workers. Initially there is no trade, but an enterprising Maxitry merchant acquires a boat and trade starts in two commodities, peaches and coconuts. It is found that to expand production of peaches by 1 ton in Maxitry, 10 tons of coconuts have to be given up, and in Minitry 15 tons of coconuts have to be given up."

(a) Which country has the comparative advantage in peach production?

Maxitry/Minitry/neither/both

(b) Which country has the comparative advantage in coconut production?

. . . .

(c) Which country is likely to be an exporter of coconuts?

. . . .

(d) Which country(s) benefits from specialisation and trade, assuming problems of adjustment to the new patterns of production can be successfully overcome?

. . . .

9.5 With a given quantity of land, labour and capital Country A can produce 12 tons of wheat or 12 tons of maize, but Country B can, with the same given quantity of resources, only produce 4 tons of wheat or 6 tons of maize. Assuming that these figures remain constant, specialisation and trade should be in such a direction that Country *A/B* exports wheat and *A/B* exports maize. (Cross out the inappropriate letters.)

9.6 In Country C and Country D the prices of beef and butter before trade starts are as follows:

	Country C	Country D
Beef	£1/lb	$1/lb
Butter	£0.33/lb	$0.5/lb

Assuming perfect competition exists in both countries:

(a) Which country will specialise in and export beef?

C/D

(b) Which country will specialise in and export butter?

C/D

9.7 Using the information given in 9.6 above, can we predict the relative prices of butter and beef after specialisation and trade have occurred?

Yes/No . . . . . . Level/Possible Range . . . .

9.8 Given that we can easily prove that some trade is better than no trade, give three reasons for tariffs or other trade regulators which could be supported by advocates of the theory of unfettered trade:

(1) . . . .

(2) . . . .

(3) . . . .

9.9 Cross out the inappropriate alternatives:

"The 'Terms of Trade' are said to 'improve' when the *value of exports/index of export prices rises/falls* relative to the *value of imports/index of import prices.* Such an 'improvement' will have an effect on the Balance of Payments which *is always beneficial/is always deleterious/can be either beneficial or deleterious depending on circumstances.*"

The following list relates to Exercises 9.10 to 9.13.

(a) Value of goods imported.
(b) Interest payments to foreigners on investments they have previously made in the UK.
(c) Spending abroad by British tourists.
(d) Value of goods exported.
(e) The investments made by British people abroad during the year.
(f) Earnings by British transport companies in shipping goods around the world.
(g) A loan to the UK from the International Monetary Fund.
(h) Spending by foreign tourists in Britain.

9.10 Which of the above are included when calculating the Balance of Payments on Current Account?

9.11 Which items would be included when calculating the Balance of Trade?

9.12 Which items are treated as Invisibles in the Current Account?

9.13 Which items form part of the Capital Account?

9.14 A country is allowing the exchange rate of its currency to "float" against other currencies. Which of the following items are likely to put downward pressure on its exchange rate?

(a) A rise in the value of exports.

(b) Domestic inflation.

(c) Investment abroad by its nationals.

(d) A rise in earnings from overseas investments owned by its nationals.

(e) A rise in the value of imports.

CHAPTER 10

Government Policy
and Agriculture

THIS book has concentrated on the underlying principles of economics which, while they are of great importance to those who wish to understand the agricultural industry, are applicable in many other contexts. While the nature of the problems which students will face in the future are not predictable in detail they can feel confident that the basic toolbox of economic concepts which has been provided here will be of substantial benefit in sorting out the causes and implications of these problems. Agriculture has been drawn on to provide examples of economic principles, but this was never intended to be a book solely about the farming industry or even the rural economy. That would be too narrow.

Nevertheless, there is a good argument for bringing together the various characteristics of agriculture which have emerged in the course of this text and to assemble them into a coherent section dealing with the problems it faces and the policies used to tackle them. This is particularly important when so much debate goes on about the costs of the action which Government (national or in the form of the collective action of the European Community) takes towards agriculture. Why Government is involved demands our attention. This means that we must examine the objectives of policy and the ways that the various actions taken meet, or fail to meet, the aims set for them.

Why Do Governments Get Involved
with the Economy?

Let us start from a general review of why governments wish to interfere with the running of the economy. Later we will move on to examine the special case for intervention in agriculture.

In market-based economies, such as the UK, the price mechanism is the main way in which the fundamental decisions of what is produced, how it is produced and who gets what is produced are settled (Chapter 1). The interaction of supply and demand has many admirable properties for this purpose (Chapter 4). It operates as a way of reflecting willingness to purchase and to produce without the need for a mass of data collection and planning. Those entrepreneurs who respond to meet consumer demands sell their output and prosper, serving the interests of both themselves and their customers. Industries where demand is growing will grow and those where demand is falling will contract.

Competition between producers means that resources go to those who can use them in the most effective way. Efficient producers grow, while the inefficient are squeezed out and new technologies which make production more efficient get taken up (Chapter 5). Comparative advantage leads to specialisation and exchange, resulting in trade both within countries and across national boundaries (Chapter 9). And because the market mechanism involves stable equilibria, such as between supply and demand, the system is largely automatic and can for the most part be left to itself.

However, a completely free market is not a perfect way of solving the basic economic problems. Flaws occur in various forms and governments will wish to intervene to modify the outcomes, "correcting" for what they perceive to be the failures of the system and achieving a better overall result in terms of the welfare of society as a whole. The main deficiencies can be classed broadly as follows:

(a) *Imperfections in the market mechanism.* We have seen that the market mechanism is not good at taking externalities into account (Chapter 7). It does not ensure that marginal social cost equates to marginal social benefit. This would lead to an inadequate supply of public goods and the external costs imposed by polluters would lack any curbs. Consumers are not always in the best position to know what is in their personal interest, as in the cases of smoking and car seat belts. The mechanism is also subject to the growth of market power by monopolies and other forms of imperfect competition. We also know that, at aggregate level, the whole economy does not automatically operate simultaneously at high levels of employment, low inflation and sustained moderate growth rates (Chapter 8). While there are some self-correcting mechanisms at work, and economists vary in their view of how strong they are, in practice some steering of the economy seems to be necessary.

(b) *failure at reflecting non-economic goals.* The market is sensitive to purchasing power, and those without this power will not have a direct impact on what gets produced and on who consumes it. The outcome of market-determined solutions may not reflect what society prefers, as reflected by the political system, including the way that people vote in elections. For example, it may be agreed to be in the national interest that young people should have equal opportunities of access to education irrespective of where they live or the type of home background. In a completely free market system, where education had to be paid for by the person receiving it, there are grounds to think that those from poorer backgrounds would be disadvantaged.

For these sorts of reasons Government will wish to intervene in the economic system. Its goals will include those listed earlier in Chapter 1, and include the maintenance of national security, stability in the currency (avoidance of both domestic inflation and erratic exchange rates), equity in terms of access to health and other basic services, and protection of the environment. In practice there are a large number of interventions going on at any one time—an array of policies with economic, social, environmental, defence and (because in a democracy governments wish to be re-elected) political ends in mind. Agriculture and the rural economy will be affected, directly or indirectly, by each of these broad categories.

The Policy Process

It is useful, before moving to policies which are of immediate concern to agriculture, to consider the process by which policy is formed and applied. This process will be of widespread application in many other situations and its study is in line with the stance of this book of examining the principles first and then choosing examples from agriculture. The process has four basic stages, shown in Fig. 10.1. These basically ask the questions: (i) Why is the policy necessary? (ii) What objective is the policy trying to achieve? (iii) What mechanism is used to bring about the desired results? (iv) What is the outcome of using the mechanism, and how does this compare with the intended effect?

Fɪɢ. 10.1   *The Policy Process*

Comment

Stage

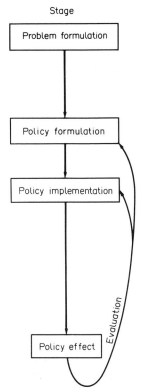

What are the problems that are to be tackled? Examples include the ensuring of a safe and secure food supply, the avoidance of environmental damage, access to the countryside for recreation, a healthy rural economy with many job opportunities.

What objectives should be set for each policy? For example, which environmental features should be conserved? How many jobs should be created in rural areas?

What alternative mechanisms for reaching the policy objectives are available? For example, are rural jobs best created by supporting the incomes of farm businesses or by giving investment grants to other types of private firms to allow them to expand? Should we use money incentives (grants, loans, tax concessions), legislation, free advice, or other means? Should schemes be locally administered or run by central government?

What is the outcome of the action taken? How do the results match up to the objectives? Would any changes seen have taken place anyway? Have data been collected to enable evaluation to take place, or can they be collected? How can the lessons learned through evaluation be fed back into the policy process in terms of the better setting of objectives and the use of more effective mechanisms?

(i) *problem formulation*. The first stage is the recognition that there is a need for some Government intervention. This may sound surprising, but the need for a policy may not be self-evident. In some cases the need for action is blindingly obvious. For example, if people are becoming ill because the nation's food supply is contaminated by inadequate health care in slaughter houses, any government will recognise that it faces a problem and will wish to do something about it. In many other cases, though, the recognition of a problem will be less immediate. For example, the awareness that something should be done about the way that farming has been encroaching on wildlife to the extent that society now wishes to protect the natural environment required the recognition that there was a danger in continuing in the old ways. Questions had to be asked, such as: "What is the real extent of the loss of hedgerows and birds?" and "Does it matter?" Society was forced to recognise that changes were taking place and to sort out its priorities.

Priorities change. In times of severe food shortage, such as in a war, environmental considerations would not count highly, but in a situation where food is plentiful and there is the opportunity to take into account the broader aims of society, a greater priority might be attached to the quality of the environment. We now think that we should attach a larger value to preserving part of the national stock of old buildings than was the case twenty years or so ago. On the other hand, the idea that everyone should have a job has been given a lower priority in the attempt to achieve other policy aims, such as the control of inflation. What is not a problem worth Government intervention at one time and in one set of circumstances is viewed as a real problem at other times, and vice versa.

In a democratic society, the process of setting the broad policy aims and achieving a balance between these aims is left to the political system, though individuals and groups can be very influential in steering the system. For example, in the case of the impact of farming practices on the environment, it needed some individuals and pressure groups to bring to the attention of the general public and the politicians the extent of the changes that were taking place. There is no guarantee that the view which predominates is necessarily the most accurate. Pressure groups often have to overstate their case in order to attract attention, and it is easy to lose perspective. Sometimes individual Government ministers become preoccupied with a particular line of argument and can exert disproportionate influence. Nevertheless, despite its drawbacks, the political system does seem to work in a way which enables real problems to be brought to the surface as a prerequisite to doing something about them.

(ii) *Policy formulation.* The next stage is to decide on the type of action which is necessary to meet the problem in hand. The responsibility for formulating these lies primarily with Government—ministers and civil servants. The critical part of policy formulation is to decide on the objectives which Government intervention is designed to achieve. These hold the key, since if they are not set out clearly and specifically it will be impossible later to judge adequately whether the actions which have been taken have turned out to be successful or not. For example, if the aim is to ensure that children living in the countryside can receive schooling without excessive travelling, an objective which could be tested might be to reduce the percentage of children living in rural areas who have to travel for more than an hour each day from, say, 20% to 10%.

(iii) *Policy implementation.* This consists of the design of programmes of action by which the objectives can be brought about and putting them into practice. There is a wide range of types of policy programme which can be used. Let us take, for example, the general policy aim of conserving areas of natural wildlife habitat in the countryside undisturbed by commercial agriculture. In pursuit of this aim the Government may determine, as a specific objective of policy, that land containing certain types of wildlife or environmental features should not be disturbed and not ploughed. What alternative types of scheme might they use? They might consider the following:

(a) passing laws on what land can be used for, with or without some compensation for the occupiers of land who find their options now limited;

(b) purchase of the land, so that the Government can directly influence what it is used for;

(c) offering farmers and owners financial incentives to encourage them to enter into management agreements not to disturb wildlife features;

(d) educating farmers and the general public on the importance of conserving the nation's environmental heritage. Part of this might involve designating certain areas as being of outstanding natural beauty.

Which method, or combination of methods, is actually used will depend on a number of factors, including the total cost, the administrative problems involved, the political acceptability of the alternatives, and the amount of value put on the wildlife to be protected. All four approaches were in use for different types of environmental protection in Britain in the 1980s.

Often there are alternative ways of administering the policy programme. For example, should they be controlled from Central Government or should local bodies, such as County Councils or National Park authorities be given the responsibility of running them and the finance to carry them out? Some programmes, such as support of prices of agricultural commodities brought about by taxing imports, can be best organised centrally but others, such as grants for encouraging farmers to repair old stone walls to make upland areas appear more attractive, are probably best left to local bodies to work out the details and organise.

(iv) *Policy effect.* If policy programmes achieve the objectives set for them, they can be considered as being effective. To return to the rural schooling example, if grants were made to local bus companies to enable them to run more services, and the percentage of children having to spend an hour or more travelling fell from 20% to 10% (or below), the objective had been reached. If, however, the percentage had not fallen at all, then the policy was not successful. If it fell to, say, 15%, the policy had been partly successful. The process of judging the outcome in relation to the objectives and the resources used is termed *evaluation*. This evaluation is rarely a simple and clear-cut process. As the name implies, more often it is a matter of subjective judgement, though taking into account as much objective information on the performance as is possible.

Evaluation has some difficult problems to face. The first is to have available adequate data on policy programmes to enable their performance to be assessed. To make this possible the decision has to be taken early in the life of a policy programme to set up the necessary administration to collect it. This systematic data collection to assist in evaluation is known as *monitoring*. Let us take as an example a Government programme to encourage elderly people to insulate the lofts of their homes by offering them 50% grants on the cost, the objective being to achieve an "x" per cent of insulated houses. The sort of information needed to evaluate this programme will include the amount of resources which have gone into the programme, such as the cost of the policy, including the administration costs. Also required will be statistics on the results of the programme, such as the number of people who applied and the number who actually went ahead with insulation. We would also want to know the number of people who were eligible to apply but did not.

Apart from data availability, evaluation has to try to determine what would have gone on in the absence of the policy programme. Continuing the insulation example, the evaluator would want to know also how many would have gone ahead without the grant anyway; if most of the elderly people intended to insulate even without the grant, the spending by the Government may have been largely wasted. The real impact of the policy is the additional amount of insulation activity which was caused; this aspect is often described as attempting to assess the programme's *additionality*. And there may be *side effects* that have to be considered. Perhaps the availability of the grant, by creating additional demand for insulation material and the services of contractors, will create extra employment and incomes through the *multiplier* effect (see Chapter 8). The workings of other policy programmes, such as special payments to the elderly in times of severe cold weather to assist with their heating costs, may be affected by the installation of the insulation—there may be fewer claims for help because houses will be warmer and therefore the public expenditure may be lower than planned. This grant scheme may also reinforce other policy programmes which aim to lower unemployment, and the full evaluation of insulation grants should take this into account. But, if the elasticity of supply of these goods and services is low, in the short term the extra demand may bid up the price of insulating a house to the extent that the net cost to the householder after receiving the grant may be little different from the original cost without the grant (Chapter 3). Some of these questions can be answered by the use of *control* areas, such as parts of the country in which grants are not offered, and by drawing comparisons with what happened before the programme was introduced. A *baseline study* is one which tries to establish what a situation is before a policy programme tries to change things. However, in many cases this is difficult or impossible to carry out.

In view of the problems, is this evaluation stage of the policy process worthwhile? The answer must be, in most cases, yes. This is because, even though clear-cut answers cannot always be given and inevitably an element of judgement is involved, attempting to undertake evaluation will expose the more blatant examples of failure or of weakness in the logic behind policy. It will lead to a questioning of why certain policies are undertaken and the weeding out of the least effective programmes. So often a policy programme, once introduced, tends to keep going without anyone asking why, or whether it is any longer necessary. Unless this examination goes on, it will be difficult to stop programmes that are ineffective or which are no longer relevant to the problems of today. Without such rationalisation, it is difficult to obtain finance for new Government policies in areas where new and pressing needs have arisen.

As we will see below, there are many examples within agricultural policy of Government programmes which were set up many years ago, when there was a critical need to expand farm output, and which have outlived their usefulness. In some cases they now positively act against the public interest. Funds used to support agricultural production, if released, could be used in many other ways within the economy, perhaps even in supporting the non-agricultural aspects of the rural economy such as in the creation of non-agricultural jobs in villages, in the support of rural services (grants to rural buses and post-offices) or in other ways of improving the welfare of rural society (subsidies to village halls?). But, naturally, the people who are currently benefiting from the present policies are reluctant to see change. As was noted in Chapter 9 in relation to the benefits from trade, it is easier for a relatively small number of individuals who would be harmed significantly by a change to band together to influence policy-makers than it is to organise large numbers who individually would benefit only a small amount, even if on balance the welfare of society would be improved by the change. It might be possible for the beneficiaries to compensate the losers, but unless the mechanism exists by which the compensation is actually paid, the vested interests may prevent any change occurring.

Government and Agriculture

In virtually all countries, whether high income and mature or developing, free market or centrally planned, the Government is in some way involved with its agricultural industry. Most of the students using this text will be mainly concerned with the market-based economies of Western countries, and these share many reasons why there are Government policies towards their farming industries. These similarities flow essentially from the economic characteristics of agricultural products and of the technology of production which are associated with the levels of income which these countries broadly share, though there are differences stemming from particular geographical, social and historical circumstances.

Let us start with the broad aims of Government policy as they affect agriculture. These can be divided into two groups. First there are the general aims of Government in which agriculture can play a role; these policies use agriculture in an *instrumental* way to achieve a policy aim. Second, there are policies specific to agriculture which are undertaken in order to counteract the economic characteristics of the agricultural industry itself. These policies are concerned with the *intrinsic* problems of agriculture. Often both sorts of aim are set out together in some official document, such as the 1957 Treaty of Rome which in its Article 39 describes those for the Common Agricultural Policy.* But such statements are not all-embracing and in practice policy aims evolve over time and are best interpreted from looking at what governments actually do.

It should also be remembered that, in an economy which consists of interrelated parts which are connected through the price mechanism, policies which are aimed primarily elsewhere will have some impact on agriculture. For example, the use of interest rates by Finance Ministers to control aggregate demand and inflation (see Chapter 8) will have an impact on farming through increasing its borrowing costs. Taxation policies designed to redistribute wealth may influence the sizes of farms; because agricultural land counts as an asset on which tax may have to be paid when wealth is passed from one generation to another,

---

* Article 39 states that the aims of the CAP are (i) to increase agricultural productivity by promoting technical progress and by ensuring the rational development of agricultural production and the optimum utilisation of the factors of production, in particular labour; (ii) thus to ensure a fair standard of living for the agricultural community, in particular by increasing the individual earnings of persons engaged in agriculture; (iii) to stabilise markets; (iv) to assure the availability of supplies; (v) to ensure that supplies reach consumers at reasonable prices.

parts of farms may have to be sold off by heirs in order to raise the sums required. In both cases governments may wish to introduce special treatment for agriculture as part of general policies. What can be called "agricultural policy" is therefore not easy to define precisely. Here we will restrict attention to those Government interventions which appear to have the agricultural industry as the main economic activity on which the policy is meant to work or which has farmers and their households as the group who are meant to benefit.

(i) Agriculture and General Policy Aims

Agriculture has an important role to play in achieving many of the wider aims which governments have for their countries. The most important of these are as follows:

(a) as a provider of a *secure food supply* for the population, assisting in the defence strategy for the country;

(b) as a source of *economic growth*, releasing resources which can be used to develop other industries;

(c) as a way of assisting the country's *trade* situation, such as by improving the balance of payments and international currency exchange rate by replacing imported food or by generating export earnings;

(d) as a way of *developing rural areas* by ensuring that jobs are available in the countryside, so that the rural economy is kept viable, and that a mixture of types of people will still live there;

(e) as a way of *protecting the environment* by maintaining a landscape which is attractive to visitors from the urban areas for recreational purposes, and which conserves the wildlife and other natural features which the population as a whole considers as part of the national heritage.

In each of these cases the intention is to provide benefits which are shared widely throughout society, not just going to the agricultural population or even those who live in rural areas. For example, the defence offered by a secure food supply, so that the country cannot be starved into submission by a potential aggressor state cutting off food supplies from abroad, is shared by the whole of society. Some of these aims, it could be argued, mainly concern the urban population. It might be thought that an attractive countryside mainly benefits those people who do not normally live there but who like to visit it.

Historically the most important of these general policies is that of ensuring a *food supply*. The reason why Britain, for example, has a separate government department concerned with agriculture (with forestry and food) can be traced back to the necessity of feeding the nation during wartime, and especially during the Second World War. During this time food production had to be expanded dramatically to compensate for supplies from abroad made impossible by hostilities. In other European countries, which suffered more directly from the fighting, the memory of starvation in the towns and cities was a vivid factor which embedded the notion of food security firmly in the minds of politicians in the post-war period. As with many Government policies, there are various ways in which food security can be brought about. One would be to keep large stockpiles of food reserves, but these may be expensive to maintain (such as the refrigeration of meat) and can only partly meet needs, and only until the reserves run out. Or some land and machinery can be kept in reserve, so that output can be expanded fairly quickly. But often governments decide that, in order to safeguard food supplies in times of emergency, they wish to have a high degree of self-sufficiency in peacetime. Often this means having an agricultural industry which is bigger than would happen if free market forces were allowed to operate freely.

In Britain, the economic recovery of the later 1940s and early 1950s needed agricultural expansion to assist the country's poor *trading position*, initially with the objective of replacing imports at almost any cost but then, once the critical period had passed and food supplies had expanded, more selectively. Even in the 1960s and 1970s there was a feeling that agricultural expansion should be encouraged if there was room to replace imports. Again, this meant having an enlarged agricultural industry.

The main way that this was achieved in the UK was to offer farmers higher prices for their products, though there were also subsidies on the factors of production which farmers used (such as subsidies on the costs of fertilisers and on buildings) and encouragements to improve the technology of production by providing a free advisory service and the Government finance of scientific research related to agriculture. As will be seen later, this support of the agricultural industry for national purposes is in the same direction as is suggested by many of the policies aimed specifically to assist the agricultural population. Consequently, many of these forms of support can still be found operating in the UK long after the food security and trade arguments have lost their force. The impact of these interventions in the free market are not difficult to predict, knowing the basic demand and supply theory described in Chapter 3. Higher prices will cause agricultural producers to expand up their supply curves, and these will be shifted to the right by lower input costs or improvements in technology. All have the effect of increasing the amount coming on to the market.

Agriculture has often made major contributions to *economic growth.* In Britain there has been constant ecnouragement of higher performance by the agricultural industry, partly to meet balance of payment goals but also because the benefits of its improved productivity spread throughout the economy. The main attention has centred on labour productivity. The Government long believed that productivity could be best achieved by encouraging investment in additional capital, and from the mid 1950s to the mid 1980s gave grants to farmers for the purchase of new buildings, drainage and other capital goods. The thinking behind the economic growth argument is basically that an expanding agricultural industry, improving its output from its productive resources through technical advances, soon finds that it has productive resources which can be more profitably employed in other activities. Because of the economic characteristics of its products, to which we will return later, resources in agriculture often have lower marginal value products than if they were used elsewhere. Consequently, governments will be on the lookout for ways in which manpower can be transferred to other industries, with an overall net benefit. This is in line with the Principle of Equimarginal Returns experienced in many sections of this book. If there are problems to the free flow of productive factors, resulting in occupational immobility (see Chapter 6), governments will wish to ease them.

Looking back in history, the rapid economic development in Russia following the Revolution in 1917 was seen to hinge on improving the performance of agriculture so that both labour and capital could be shifted to the heavy industries where rapid expansion was needed. The reorganisation of farming along large-scale lines, with collective farms, and the forced relocation of labour, made possible by central economic planning and an oppressive political regime, was part of this. More recently, a major contribution to the rapid growth of many continental countries of the European Community in the first two decades following its establishment in 1957 was the substantial transfer of labour out of agriculture to other industries in which it could be more productive.

This source of growth was largely denied to Britain, where there was a far smaller percentage of the population engaged in farming and where output per man in agriculture was already relatively high by the start of this period.

In the 1980s the higher levels of unemployment in the economies of many EC Member States have thrown attention on the potential of agriculture to retain employment. It has also been seen as one channel through which the problems of the rural economy could be tackled. Often these centre on the need to keep a viable size of population, which then can sustain rural services (schools, shops etc.). Some remoter parts of the EC are facing severe depopulation problems as farming declines. Though non-agricultural industries can be encouraged, in practice an easier solution, at least for the short term, is to support farming there. Agriculture is thus being used as a tool of social policy—an *instrumental* role. A very similar situation exists in the upland areas of Britain, though the threat is less one of depopulation than one where there is a change in composition of the sorts of people found there, with more holiday homes, retired people and professionals from towns, and less opportunity for lower income families to live and work in the area.

The last item in this group of general policy objectives which relate to agriculture is the environment issue. Although farms are often recognised as multi-product firms, which carry the necessity of considering the balance of outputs and the use of a range of inputs (see Chapter 5), thinking is often restricted to the sorts of commodities that might be classed as food for humans or animals, possibly widening to forestry products and some industrial products, such as flax. However, governments recognise that agriculture also produces non-tangible outputs which are valued by the rest of society. Some, like tourist services on farms, fall into the category of "private" goods and can be handled by the market mechanism, though farmers may need to be made aware of the potential and trained to exploit it. However, there is now an enhanced sensitivity among governments in many industrialised countries, and especially the UK, to the importance that the public attaches to the appearance of the countryside, and a willingness to pay for the conservation of its special features. This landscape/conservation character can be regarded as another output from the agricultural industry, since the land-using activities of farmers and others determine what the countryside looks like and the wildlife features it contains. Agriculture is providing a service for the non-agriculturalist, through recreation or just the well-being of knowing that the countryside is "there" and being looked after, which is valued in the same way as its food-security role. Because of the "public good" aspects of such features, Government intervention is necessary to enable the wishes of society to be implemented, even if only partially. Reflecting the greater affluence of the population in general, we can expect to witness an increasing role for policies which encourage farmers to use their land in ways which are consonant with the emerging ideas of what the countryside should look like and the activities which should take place there.

(ii) Policies Designed Specifically for Agriculture

The second group of reasons why governments have policies for their agricultural industries stem from the economic characteristics of agriculture itself. First we must summarise these characteristics which have been mentioned in previous chapters. For simplicity we assume that food is the only output from agriculture.

(a) Basic food products typically have low price elasticities of demand.

(b) Basic food products typically have low income elasticities of demand, so that little of any extra income is spent on the output of agriculture.

(c) Agricultural production is subject to unpredictable variations from year to year, mainly caused by weather. The result of this, and the low price elasticity of demand, means that the revenues from sales by farmers vary greatly.

(d) The industry is composed of an atomistic structure which approximates to perfect competition among producers. Though there is great diversity in terms of farm size, even the largest are unable to have a significant effect on the overall market price. There is a possibility of suffering from a poor bargaining position when dealing with large suppliers of agricultural inputs or buyers of farm commodities. This can be avoided to some extent by forming farmer-controlled co-operatives to buy and sell. The bigger problem of this structure, though, is the conflict it produces between what is best for the individual and what is best for farmers as a group. Uncoordinated production decisions are also conducive to the establishment of price cycles.

(e) As technological improvements are taken up by individual farmers, starting with a few innovators but spreading to the majority in time, the industry is characterised by a gradually rising output. This expansion, when faced by an inelastic demand, has the effect of pushing down agricultural prices over the long term. The structure of the industry means that it is in the interest of individual producers to expand even though they know that it is not necessarily in the interest of all producers.

(f) Falling prices squeeze incomes both of the industry as a whole and at the farm level. Though the volume of output may increase over time, the income left to reward its producers is less. Farmers look for ways of countering this, such as further expansion of area, replacing labour with machinery where this enables costs to be cut, and diversifying into non-agricultural activities. The smallest farms can no longer provide an adequate reward for their operators and are in the weakest position when competing with other farms for extra land. Many of them have to give up. Their land is taken over by other farmers, usually the medium or large operators, to increase their farm size.

The combination of the above factors means that agriculture is a declining industry in a country such as the UK. Technological advance which generates extra output, and which spreads through the industry because it is in the interest of individual farmers to adopt it, ultimately leads to people leaving farming, the disappearnce of the small farm and increasing average farm sizes. Over time the minimum size of a farm capable of generating an adequate total income is bound to rise. These changes are a reflection that the price mechanism is performing its function of adjusting the economy to the new state of technology. The benefits of improved ways of producing agricultural commodities are being passed on to consumers in the form of lower prices, and productive resources which are no longer needed in farming are being released to be used in other parts of the economy. In national terms these changes are reflected in a persistent fall in the total number of people engaged in agriculture and a shrinkage in its share of the national labour force and of the proportion of the national income generated by agriculture.

But there are additional factors which must also be considered. Continuing the sequence:

(g) Farms are typically small businesses which are run mainly by families, often with the family providing much of the labour, and with the farm providing a place to live as well as a way of making a living. Personal and business wealth is closely mixed. Thus when small farms are no longer viable, there will be social problems as well as business ones, such as the need to find a new place to live as well as a new occupation for the farmer and family. Often this means moving to an urban area, which may not be something which comes naturally to a person used to living in the countryside.

(h) The transfer of resources out of agriculture, especially labour, is often difficult and slow. The occupiers of small, unviable, uncompetitive farms tend to be elderly and with little experience of other ways of earning a living. There may well be an emotional attachment to the land which makes them reluctant to leave. Whatever the cause, they suffer from occupational (and often geographical) immobility. Rather than transfer to other occupations, small farmers may struggle on, perhaps inefficient in their use of factors of production but nevertheless helping to generate the output which in turn keeps product prices low.

(i) In the process of reducing labour in agriculture, the first reductions tend to be made in the hired labour force rather than in the farmer's own family. These are often the least able to acquire housing and therefore jobs in new locations.

(j) Both among independent farmers and in the hired labour force, the most likely to shift to non-agricultural occupations are the younger and better-educated. This means that agriculture may contain a disproportionately large number of elderly people, especially in the more remote rural areas where leaving agriculture means moving away to a town job. Changes in the composition of the rural society may carry implications for the viability of rural communities.

Taken together, it is hardly surprising that there are cries for the Government to do something to counter the pressures on farming brought about by the fundamental economic forces at work. Few people relish change, especially the sorts of structural changes which are necessary to accommodate the technical advances which are going on. The inevitability of the changes does not stop demands coming from farmers and their representatives for public support for farm product prices, which they see always falling in real terms and in relation to the costs which farmers face. The hardship caused to the small farmer is pointed out, often by the larger farmers who stand to benefit most if the Government accedes to demands for support. The threat to rural communities is highlighted. The need for a secure food supply and an attractive countryside is emphasised. Each assertion and claim for support should be examined with care. Is the logic sound? Are there alternatives to the support of agriculture which could achieve the ends more effectively?

From our analysis of the characteristics of the agricultural industry we can conclude the following:

(a) the Government may wish, as a policy objective, to *stabilise the markets* for agricultural commodities and to prevent the damaging effects of large random changes of income from year to year. Also relating to markets, there may be the need for the Government to keep a watch on the possibility of farmers being exploited by monopolists and monopsonists, and to do this by the encouragement of farmer-controlled co-operatives and marketing agencies;

(b) that a policy may be necessary to assist *structural adjustment* in agriculture, by helping some farmers to leave the industry, to help others increase their farm sizes to viable proportions, to train farmers in alternative types of economic activity (such as the setting up of craft industries using farm buildings or the development of tourist facilities on farms to give additional income sources);

(c) there is the possibility that, among some sectors—typically among small farms where resources are immobile so that structural adjustment cannot take place with sufficient speed—incomes will be so reduced that the farmers and their families cannot enjoy a standard of living which the rest of society would consider a bare minimum.

Governments may set as a policy objective *income support* so that the agricultural population should have a "fair" standard of living. It should not go unnoticed that the last two objectives are potentially in conflict with each other. Structural adjustment in agriculture is ultimately a necessity, but it can be fended off for a time if governments are willing to support farmers' incomes.

Agricultural Policy in Practice

Each of the above aims of policy should properly be considered separately. It could well be that, for example, income support would be best achieved using policy programmes which are quite separate from those which aim to improve productivity. But in practice many of the Government and EC policy programmes were set up in the post-war period when it was thought that several, if not all, the aims could be achieved with the same types of intervention in the markets. For example, if the Government were to raise prices for farmers in order to induce them to produce more and to back this up with grants to encourage investment in extra machinery to increase productivity, this would simultaneously improve the incomes from farming and, assuming that the country were a food importer, lower the import bill and assist the balance of payments. When the belief predominated that prosperous farming produced an attractive countryside, and that farmers would only protect wildlife if they had enough income to allow them to leave parts of their fields undisturbed, support of farm product prices would also realise these environmental goals.

The ability to achieve all these aims of policy mainly through a single mechanism—support of farm product prices—is no longer seen to be possible. Price support leads to substantial conflicts between agricultural and environmental goals; farmers are encouraged to use their land more intensively and to bring into cultivation marginal land which for environmental purposes should be left as rough grazing. Europe no longer wants additional output, and other ways must be found of achieving income aid for farmers; furthermore, the support of prices means that most of the benefit goes to the larger farmer whose living standards are likely to be good and the low income farmer on the smallholding at the margin of viability gets relatively little help. Even so, price support by blunting the spurt of economic hardship also works directly against any policies which are aimed at improving the size structure of agriculture by helping the operators of too-small farms to leave the industry. This may store up even greater policy problems for the future when the costs of product price support rise to levels which the Government is not willing to face.

The implications of the conflicts between policy objectives have not been fully reflected in the way that governments attempt to manipulate the agricultural industry. Without doubt the most important agricultural policy programmes in the UK, the European Community and the USA, both in terms of public money spent and in their degree of influence on the size and shape of farming and the countryside, are still those which support the prices which farmers receive. They deserve special study.

The Support of Farm Commodity Prices

There are two main types of mechanism which have been used to raise the prices of agricultural commodities above the free-market level and therefore the size of the home agricultural industry and the incomes of farmers. The first is a system of guaranteed prices, with the Government making up the difference between the market price received by the farmer for his output and the guaranteed level by means of a direct payment, called a "deficiency payment". Fig. 10.2 illustrates a simple system. The demand curve represents the demand from the UK population for this food material. The curve Sh is supply curve from domestic producers of this food. If there was no supply from abroad, the market price would be Pd. However, if the commodity can be imported without restriction the price that the UK population will be willing to pay will be determined by the price at which imports can be obtained. This is represented by the horizontal supply curve labelled Sw. We assume that UK importers can buy as much as they wish without affecting world prices. Thus, in a free market situation, the price of this food in the UK (Pfm) would be the same as the world price (ignoring transport costs). At this price UK farmers would supply quantity qh1; consumers would buy qt1 and the difference would be imported.

If the Government decides to guarantee that UK producers receive a higher price, labelled Pg in Fig. 10.3, home producers will expand up their supply curves to output qh2. The price which consumers pay is

FIG. 10.2   *Supply and Demand without Government Intervention*

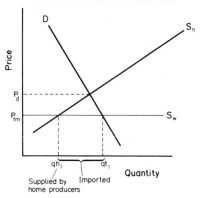

Note: the demand and supply curves are drawn with these slopes for illustrative purposes. Demand is shown to be relatively inelastic, and supply relatively elastic. While these generalities hold, the slopes and positions of the curves have been chosen to make points and may therefore be rather extreme. This also applies to other figures in this chapter.

FIG. 10.3   *The Effect of a Guaranteed Price and Deficiency Payment*

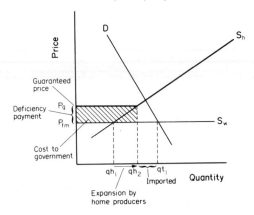

not affected, so the quantity they buy remains the same. The net effect is that less is imported from abroad. The bill to the Government, which has to be financed from taxation, is the shaded area on Fig. 10.3. It may be somewhat larger than they anticipated, because UK farmers have expanded their output. Also, if world market prices fall the amount of deficiency payment will have to be increased; changes in import prices are a source of uncertainty which Finance Ministers may not welcome. Because poor people generally pay little tax, the burden of this support system is not felt by low income earners. It might be noted that whatever happens to market prices, producers will always get the guaranteed price, the deficiency payment making up any difference. In effect, domestic producers now have a new market supply curve which becomes completely inelastic at prices below the guarantee. Also important to note is that, as the guaranteed price is set at higher and higher levels, the amount of imports is reduced, eventually to nothing. If the guarantee is set above the price which would balance demand with domestic supply, there will be a surplus which will be exported.

The other main method used to increase prices for farmers is to tax imports. This is shown in Fig. 10.4. The effect is to raise the supply curve for imports from the rest of the world by the extent of the tax. Unlike the deficiency payment system above, this results in consumers facing higher prices in the home market for their food. Home farmers expand, encouraged by higher prices, and imports are again reduced, but part of this reduction comes from a fall in demand caused by the higher prices. Governments, rather than having to fund support, find themselves receiving revenue from the taxes on imports. But the burden of the system falls disproportionately on the poorer consumers for whom food spending is important and who face the same price rises as richer consumers; the proportion of their expenditure which goes on food is greater than among better-off people.

But taxing imports can only raise the prices which domestic farmers receive if there are some imports to tax. At higher levels of taxation imports are progressively reduced. The greatest price which can be generated this way is when the tax reduces imports to zero; the domestic price would solely be determined by the interaction of demand with domestic supply. To go higher would need other mechanisms, such as the Government deciding to buy from the market to keep prices up. The

FIG. 10.4 *A Tax on Imports*

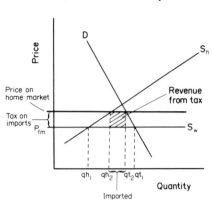

mechanism of this type of support buying is shown in Fig. 10.5. In effect, the Government is creating a new demand curve which is infinitely elastic at the price at which Government decides to intervene and guarantee to farmers. However much farmers produce, the Government will have to buy it. If technological advances in agriculture go on, the supply curve will shift to the right and the amount that has to be bought with public funds increases. As the diagram shows, the higher price chokes back the demand from other consumers, so the amount of public buying is probably more than was originally intended. This reduction will depend on the steepness of the market demand curve.

One additional problem of support buying is that the Government will have a quantity of farm produce on its hands. It can store it, and the CAP has at times built up large amounts of commodities in public storage facilities (and in private storage paid for out of public funds). There has been much comment about the grain, beef and butter "mountains" and the wine "lake" created through support buying. But stocks cannot continue accumulating for ever. Storage is expensive, especially for frozen meat and dairy products. Ultimately they must be dispersed. Various possibilities arise; butter stocks can be destroyed or turned into animal feed or used for industrial purposes. Wine can be turned into industrial alchohol.

Fig. 10.5   *Support Buying*

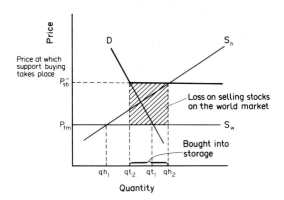

Notes: (i)   the price on the market is raised to P_{s_b} by support buying
   (ii)   consumers buy less—qt_2 rather than qt_1—because of the higher price
   (iii)   domestic producers expand from qh_1 to qh_2
   (iv)   the Government buys into storage qh_2 minus qt_2, that is total production less the amount that consumers buy
   (v)   to dispose of its purchases on the world market the Government will make a loss (buying-in price less the world price, times the amount sold)
   (vi)   support buying only works if imports are controlled, by taxes or quotas or other means.

Many of these methods are politically unacceptable—voters do not like to see good food destroyed. Some can be given away as food aid to poor countries. But the way most often used is to sell it to exporters who are given a subsidy to enable them to sell it at a loss on the world market. This also causes problems, as this subsidised produce may push down the world price, disrupting the pattern of international trade and leading to claims of dumping by countries who find their exports no longer competitive (see Chapter 9). To prevent this sort of ill-feeling, sometimes special deals are arranged with specific countries. But these can also cause political problems at home; at a time when the USSR was out of international favour because of its treatment of civil rights it was embarrassing to find that Russia was buying subsidised butter from the CAP at a small fraction of the price that domestic consumers were facing. In practice it seems that agricultural policy has been shaped primarily by what governments wished to achieve for their home agricultures, without thinking too much about the international consequences.

A further type of intervention which must be considered is the quota—a restriction on the quantity entering the market. Quotas can be placed on imports to keep prices up, but the most notable use in agricultural policy today is on domestic supplies, all imports being prevented (in effect, a zero quota on imports). Quotas on milk sales is the best-known example. Here we are concerned with the market impact of quotas rather than that at the farm level. The effect of a quota is shown in Fig. 10.6. Whatever the market price, the amount which farmers can supply to the market is limited to the quota amount. This creates a new supply curve. Consumers, faced with this smaller quantity than would be available at the equilibrium between demand and supply in an unrestrained market, have to pay a higher price. As with the import tax, this higher price poses a bigger burden to the poor consumer than the higher income one. Farmers as a group get a higher total amount of money from their sales, even though quantity is less, because prices are forced up proportionally more. How this benefit is distributed among farmers will depend on how the quota is allocated among them, but if cuts in output are applied on a percentage basis, the high output farmer will benefit in absolute terms more than the lower output producer, as with any other price-raising mechanism. Quotas only work if demand is inelastic, and the steeper the demand curve the bigger the

FIG. 10.6   *A Quota on Home Production*

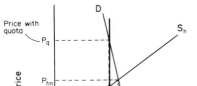
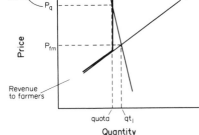

Notes: (i)   it is assumed that there are no imports
      (ii)   the rectangle showing revenue to farmers
             (price times quantity) is greater after the
             imposition of the quota.

price rise which a given quota restriction will cause. As well as the impact on consumers and the lessening of the drive for efficient production and the slow down of structural adjustment caused by higher prices, quotas have additional problems. They need policing, since it will be in the interest of individual farmers to produce more at the higher prices than their quotas allow. Pressures for greater efficiency are further diminished by the knowledge that the efficient producer, however successful he becomes, cannot compete with less efficient farmers and expand beyond his quota limit. Quotas, unless made transferable between producers by sale or other means, are said to make rigid (or *ossify*, from the word for bone) the pattern of production of the commodity to which they apply.

The Future Nature of Agricultural Policy

The policy process and its evaluation was shown earlier as being concerned with the setting of objectives and the design of programmes to achieve them. Whether or not policy was successful depended on comparing the outcome with these objectives. However, a major feature of agricultural policy is that the objectives have rarely been defined precisely and the ways in which the various aims might conflict with or compliment each other have not been thought through. Individual support programmes have usually been introduced to meet isolated problems without fully appreciating the ways that they could affect other programmes. This in part is the result of the way that responsibilities are divided. Policies for agricultural production are usually under the control of one Government department and those for the environment under another. Even within these broad areas there may be divisions; in the UK some parts of agricultural policy are largely national in design and implementation while other parts originate from the Common Agricultural Policy which is shaped by decisions taken by the EC organisations in which the UK voice is one among many.

In the 1990s some of the more obvious failings of agricultural policy are likely to be made good, at least partially. Among the changes which can be expected are the following:

(a) a recognition of the basic supply and demand characteristics of agriculture, including the acceptance that this industry will continue to experience a decline in the prices it receives for its products compared with those of other sectors and a sustained squeeze on the incomes of agricultural producers. As a result, it is inevitable that the structure of agricultural production will have to change. In part this will be met by the encouragement of part-time farming and by farm businesses diversifying into non-agricultural activities (tourism, food processing and so on), which governments will try to encourage. Nevertheless, more small farms will have to disappear and the resources they use deployed in other activities;

(b) a recognition that policies which attempt to support agricultural prices will, by delaying the necessary adjustments to the level of output and the quantities of resources retained in agriculture, necessitate even larger and more painful changes later. This will lead to a reduction in the amount of policy concerned with interfering in the market for farm commodities;

(c) a recognition that the policy objectives involving agriculture are varied and that they need to be clarified. At present there is a mixing of agricultural production objectives, environmental objectives and rural social objectives, trade objectives and others. Once objectives become clearer, the appropriate types of policy programmes can be introduced. For example, to achieve environmental objectives specific payments for farmers to manage their land to produce herb-rich meadows, or whatever, can be devised. Where the intention is to support incomes, direct income payments to those who satisfy the criteria for assistance can be introduced;

(d) a recognition that, especially in those policies where agriculture is used as a means to an end, alternative ways of reaching the objective may be preferable. The imposition of planning control on agricultural land use may be a more effective way of securing the preservation of areas of unique wildlife than the use of voluntary management agreements (with financial inducements) to farmers who own these sites. To take another example, though support to agriculture in hill areas of the UK is given in order, among other things, to keep people living there, more efficient ways of doing this may be available, such as by supporting rural factories or hotels. This will require the estimation of income and employment multipliers (see Chapter 8) for each type of public spending and in each type of location;

(e) a recognition that agricultural policies, designed with domestic aims in mind, can have substantial consequences for international trade, not only in food products but, because of the reciprocal nature of trade, in non-food products as well. Hence there will be a search for ways of achieving internal objectives which do not have harmful trade implications, and for the scaling down of those which currently lead to trade problems.

Many of these changes are already under way. All change involves risks, costs and disturbance of various forms, and bureaucracies (Government departments, the European Commission and so on) will tend to prefer to keep doing what they are already doing. Reform of policy has to overcome administrative inertia and vested interests. The speed with which they take place is in large part determined by the costs of keeping on with the present arrangements. If, for example, the costs of the current CAP, dominated by support of prices for farm commodities, threaten to bankrupt the European Community budget and perhaps even the continued existence of the Community, action could be swift. However, without this spur change is unlikely; there are many people, including the large farmers and some administrators, whose interest would be harmed by reform of the CAP and who would oppose change.

The sorts of improvements to agricultural policy envisaged illustrate the sort of questioning that should be going on all the time about the purposes for which Government is involved with the agricultural industry and the most effective way of using public funds. They indicate a greater awareness of the principles of economics, especially the concern that resources are scarce, that they require allocating between alternative uses, and to do so involves thinking about the objectives behind their allocation. Thus the study of agricultural policy fits well into the general definition of economics which emerged in Chapter 1.

Exercise on Material in Chapter 10

10.1 Place the following steps of the policy process in the correct order. Add numbers (1-4) in the brackets.

policy effect (   ); policy implementation (   );
problem formulation (   ); policy formulation(   ).

10.2 Describe what is meant by each of the following:

(a) evaluation ....

(b) monitoring ....

(c) base-line study ....

(d) side-effects ....

10.3 The Government has the aim of reducing the amount of contamination of rivers by silage effluent. Describe three types of policy intervention that might be considered.

(a) ....

(b) ....

(c) ....

10.4 Explain what is meant when it is said that agriculture is being used in an instrumental role to achieve a policy objective.

....

....

10.5 Give five reasons why a government may wish to have an agricultural policy.

(a) ....

(b) ....

(c) ....

(d) ....

(e) ....

10.6 The Government has decided to increase the price that farmers receive for wheat in the UK by buying from the market. There are no imports to consider (assume they are banned). Using the diagram below answer the following:

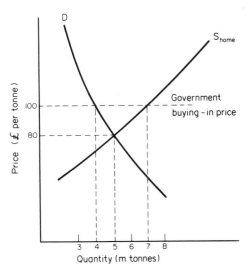

(a) If the Government buys at £100 per tonne, how many tonnes will it buy? ....

(b) How much will it spend? ....

(c) If it wants to dispose of its purchases on the world market at £50 per tonne, what will the loss be? ....

(d) What is the effect on the output of UK wheat by introducing the supported price? ....

10.7 Continuing the same example, what would you expect to happen to the following? Indicate the direction of change (up/down or unchanged).

(a) The output of other cereals? ....

(b) The income of wheat producers? ....

(c) The demand curve of the users of wheat? ....

(d) The amount of wheat that is bought by users (not counting the purchases by the Government)? ....

(e) The profits of firms who use wheat? ....

(f) The prices of the consumer products (bread and so on) which have wheat as an ingredient? ....

(g) The price of land which can grow wheat? ....

(h) The demand for chemicals specific to wheat-growing? ....

(i) The price of these chemicals if their supply is
   (1) completely elastic ....

   (2) less than completely elastic? ....

10.8 The following diagram illustates the imposition of a tax on imports of an agricultural commodity which can also be grown in the UK.

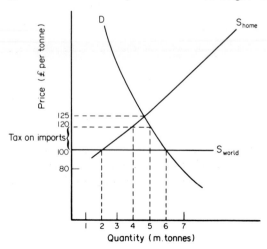

Answer the following questions.

(i) Before the tax is imposed how much is:

   (a) Produced in the UK? . . . .

   (b) Consumed in the UK? . . . .

   (c) Imported into the UK? . . . .

   (d) How much in total is spent on imports? . . . .

(ii) After the tax is imposed how much is;

   (e) Produced in the UK? . . . .

   (f) Consumed in the UK? . . . .

   (g) Imported into the UK? . . . .

   (h) Spent on imports? . . . .

   (i) Raised in taxation? . . . .

10.9 Continuing the above example:

   (a) What is the highest price that can be arranged for farmers by taxing imports? . . . .

   (b) What is the level of tax (per tonne) which just brings this about? . . . .

   (c) What is the revenue to the Government from this tax? . . . .

10.10 The following diagram illustrates the use of a guaranteed price and deficiency payment system for supporting the price of an agricultural commodity.

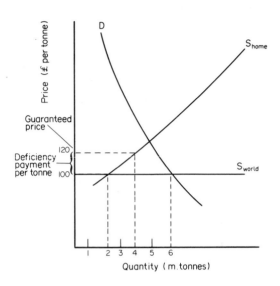

From this diagram, answer the following questions:

(a) What is the size of the deficiency payment that the Government
has to pay per tonne? . . . .

(b) What is the total quantity that UK producers supply to the
market before the introduction of the payment?

. . . .

(c) How much do they supply when the payment is introduced?

. . . .

(d) By how much does the quantity demanded from the market
change? . . . .

(e) By how much does the amount imported change?

. . . .

Suggested Further Reading

A. General texts to supplement all chapters of this book.

Begg, D., Fischer, S. and Dornbusch, R. *Economics.* McGraw-Hill, British edition, 1984.

Lipsey, R. G. and Harbury, C. *First Principles of Economics.* Weidenfeld and Nicolson, 1983.

Samuelson, P. A. and Nordhause, W. D. *Economics.* McGraw-Hill, 12th edition, 1985.

Bannock, G., Baxter, R. E. and Rees, R. *The Penguin Dictionary of Economics.* Penguin, 3rd edition, 1984.

B. Texts specific to the economics of agriculture.

Colman, D. and Young, T. *Principles of Agricultural Economics.* Cambridge University Press, 1989.

Hill, Berkeley and Ray, D. *Economics for Agriculture: Food, Farming and the Rural Economy.* Macmillan, 1987.

Hill, B. E. and Ingersent, K. A. *An Economic Analysis of Agriculture.* Heinemann, 2nd edition, 1982.

Hallett, B. *Economics of Agricultural Policy.* Blackwell, 2nd edition, 1981.

Newby, H. *Green and Pleasant Land?* Gower, revised edition, 1985.

Ritson, C. *Agricultural Economics: Principles and Policy.* Granada, 1977.

C. Farm production economics.

Barnard, C. S. and Nix, J. *Farm Planning and Control.* Cambridge University Press, 2nd edition, 1979.

Bishop, C. E. and Toussaint, W. D. *Introduction to Agricultural Economic Analysis.* Wiley, 1958.

Essay Questions

THE number before the question indicates the chapter which is of particular relevance.

1a. What is an economic problem and how does it arise?

1b. "Even in a world of plenty all commodities are scarce." Comment on this statement.

1c. What do you understand by "opportunity cost"? Illustrate your answer using choices individuals have to make in allocating their resources, and the way in which the State spends our money.

1d. "Facts by themselves are dumb; before they will tell us anything we have to arrange them, and the arrangement is a theory." Describe the process of theory formation, testing and application, with examples drawn from economics.

1e. Discuss the proposition that economics is concerned with problems of rational choice between alternative possibilities.

1f. Economics has been described as being a social science. By comparison with natural sciences, how appropriate is this description?

2a. "A consumer is in equilibrium when he is getting most for his money." Comment on the accuracy of this statement as a definition of consumer behaviour.

2b. Illustrate, with reference to indifference curves, that economics is concerned with problems of rational choice between alternative possibilities.

3a. Why is the concept of demand elasticity an essential part of economic theory?

3b. What explanations might be offered if more of a commodity is brought after an increase in price?

3c. "The success of British agriculture has benefited everyone but the farmer." Explain the background of this quotation in the light of demand and supply theory.

3d. This notice appeared in a University handout to post-graduate students. "Post-graduates are advised to apply early if they require one of the University's furnished flats. The demand for these flats far exceeds the number available." Comment on that establishment's method of allocating accommodation.

3e. Explain what is meant by the income elasticity of demand for a commodity and why it tends to differ for different commodities and different income levels. What is the significance to producers and consumers of the income elasticity of demand being (a) greater than one; (b) negative?

3f. What are the consequences for agriculture of
(i) the low price-elasticity of demand; and
(ii) the low income-elasticity of demand for most agricultural products?

3g. Discuss the implications of Engel's Law for the agricultural sector of a growing economy.

4a. Explain why fluctuations occur in market prices.
How do fluctuations reflect the degree of competition in an industry?

4b. In an industry which is virtually perfectly competitive, such as milk production in the UK, increases in output by individual farmers in their quest for greater profits may be against the interests of the industry as a whole. Explain how this conflict arises, and what measures the industry can adopt to protect itself.

4c. How does monopoly power arise? Is the existence of a monopoly necessarily something which is to be deplored?

4d. What do you understand by price discrimination? In what circumstances is it (a) practicable; and (b) profitable?

4e. Discuss the factors which a monopolist takes into account when maximising his profits. Is monopoly necessarily a bad thing?

5a. Discuss the importance of the concept of the "margin" in production economics.

5b. What are the important production relationships which an entrepreneur must optimise if he wishes to generate the highest profit? What difficulties might he be expected to come across in aiming for this goal?

5c. Economics often assume that the motivation of firms is profit maximisation. Is this assumption borne out by the behaviour of firms? What other motives might bear upon their actions?

5d. Describe the economies which can be achieved through large-scale production. In view of these economies, how is it that in agriculture, small firms can still flourish side by side with large ones?

5e. What is the relevance of time to production?

5f. Define the Law of Diminishing Marginal Returns. Discuss whether a producer should cease to expand production once his marginal returns start to diminish. Does this law apply particularly to farmers?

6a. "Profits tend to equality." Point out the ambiguities in this statement and submit a clear statement about the relations between profits in different industries and enterprises.

6b. Compare the concept of transfer earnings and economic rent with that of normal and pure profits.

6c. Discuss the role of the entrepreneur in the process of production.

6d. Define the following terms and consider the relations between them: Wealth, Capital, Land.

6e. How is factor mobility related to economic rent?
Examine the proposition that economic rent is likely to constitute part of the reward of *any* factor of production.

6f. Robinson Crusoe's fishing would be far more effective if he were to use a net. Discuss his decision to construct a net in relation to the economic theory associated with the factors of production.

7a. "The market mechanism would work very well if governments would leave it alone." Comment on the validity of this statement.

7b. Describe what is meant by "externalities". By reference to agriculture, illustrate why they should be taken into account in decisions aimed at improving the way that the economy is organised.

7c. What is a "pure public good"? Why would a *laissez-faire* economic system be expected to have a smaller quantity of public goods than everyone would agree is desirable?

7d. "Governments should pay for the repair of ancient farm buildings and country churches. They are part of the national heritage and are public goods." Comment.

7e. Discuss the alternative ways in which the external costs of silage effluent can be "internalised". Put forward an argued case for the alternative you prefer.

7f. What are the economic arguments for compelling school children to stay in full-time education to the age of 16 years when many would wish to leave at 14 years? Imperfect knowledge and externalities should find a place in your answer.

8a. Discuss the proposition that saving is a private virtue but a public vice.

8b. In what ways can the government of a developed country influence the working of its general economy?

8c. "The multiplier is a two-edge sword. It cuts for you or against you." Explain and discuss this statement.

8d. Discuss the chief factors which will cause a rise in general prices. How are such price changes brought under control?

8e. Discuss the relative merits of economic growth.

8f. Outline the problems encountered when a government of a Western developed economy attempts to achieve simultaneously low employment, stable prices and economic growth.

8g. Outline the factors on which a nation's standard of living depends.

9a. "It is comparative advantage, not absolute advantage, which determines the pattern of trade between countries." Elucidate and discuss.

9b. Although the law of comparative advantage shows that, given certain assumptions, free trade maximises welfare, most nations in fact impose tariffs. What reasons are usually advanced in support of this policy, and to what extent are they justified?

9c. Should the United Kingdom import some goods which it could produce at home?

9d. Is it economically sound to appeal to the inhabitants of the United Kingdom to "Buy British"?

9e. "Most arguments for protection for a home industry against competition from abroad are simply rationalisations for special benefits to particular pressure groups." Discuss.

9f. The import of fresh liquid milk into the UK for either liquid consumption or for manufacture into dairy products is restricted. What arguments might be put forward for and against such a constraint on trade?

9g. What economic effects might be expected to flow from the purchase by foreigners of farms in the UK?

9h. Why should a government be interested in the international exchange rate of its country's currency?

10a. What is meant by the "policy process"? Illustrate your answer with examples drawn from UK agricultural policy.

10b. What are the chief economic characteristics of agriculture in the UK? What problems do they pose for Government?

10c. What are the objectives of the Common Agricultural Policy of the EC, as set out in 1957? To what extent had they been achieved by 1990?

10d. Why are the incomes of small farmers likely to pose a problem in the EC? What options are available to deal with them?

10e. Compare and contrast the use of deficiency payments and import levies as a way of raising the incomes of cereal farmers in the EC.

10f. One objective of agricultural policy in the UK up to 1980 was the improvement of productivity. Is this objective outdated in the 1990s when surpluses of farm commodities are a major problem?

Suggested Answers and Explanations for the Exercises following each Chapter

1.1 The three missing essentials are:

(a) Resources available to man are *scarce*.
(b) Their allocation between alternative uses requires *choice*.
(c) This allocation is designed to achieve certain *objectives*.

A better definition would be "Economics is the study of how individuals and society distribute scarce resources between alternative uses in pursuit of given objectives".

1.2 The resources of land, labour, capital and entrepreneurship available to man for the production of goods and services to satisfy his wants are limited in supply. Because they are of limited quantity man has to choose how he allocates them between alternative uses. Any good which is limited in supply relative to the wants for it i.e. where choice of use is involved, is economically scarce. Any commodity which is not limited in supply is not economically scarce. No problems of allocation arise; sufficient quantities are available to satisfy all wants completely. Such commodities are termed "Free Goods".

Example:

Air: Only in exceptional circumstances is air limited in supply and hence is "scarce" e.g. in mines, submarines.

Water: Water is normally a scarce commodity. We therefore encourage people to turn off taps rather than leave them running. Prices are used here; householders pay for water through water rates and the indiscriminate use of water would be reflected back eventually in higher costs. In times of drought, particular care has to be exercised in the uses to which water is put; it becomes a more scarce commodity and other methods for its allocation may need to be used (e.g. water rationing). To a man adrift on a raft on a fresh-water lake water is not a scarce commodity; the problem of allocation between different uses does not arise. There is enough for all the uses the man can devise for water.

Sand: To the UK economy sand is a scarce resource—it is limited in supply in comparison to the requirements it could serve. Prices are used as a method of allocating it. A building firm in a sandy desert could use as much as it wished.

1.3 Positive.

1.4 Positive statements concern what is, was or will be, and normative statements concern what ought to be. Disagreements over positive statements can be settled by an appeal to the facts, while disagreements over normative statements cannot be settled in this way. Normative statements depend on value judgements.

1.5 Opportunity cost is the cost of making a choice in terms of fore-gone alternatives. More specifically the opportunity cost of using something in a particular way is the benefit foregone by not using it in its *best alternative use*. Given that the State can spend money on a hospital *or* a road, the opportunity cost of building the road is the hospital.

1.6 Which is foregone in each case?

(a) his evening of study
(b) his evening of table-tennis.

1.7 (a) the £1,000/year job plus convenience
(b) the £1,100/year job.

Alternatively

(a) the £1,000/year job
(b) the £1,100/year job plus necessary travel.

It helps if alternative choices are thought of in terms of the satis-faction which they can give. In this example satisfaction derives not only from monetary reward but also from convenience.

1.8 In terms of foregone alternatives, the opportunity cost of growing wheat is the net revenue from potatoes. If the opportunity cost of growing wheat were greater than the net revenue produced by growing wheat, the farmer, if motivated by profit, should switch from growing wheat to growing potatoes.

1.9

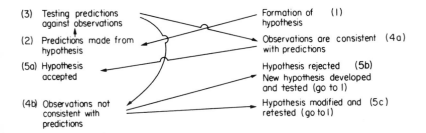

1.10a (a) Form an hypothesis. "Farmers on dairy farms of 100-150 acres in Devon who have invested in buildings over the last ten years get lower net incomes now than those who did not."
   (b) Make a prediction; "if farms were visited one would find lower net incomes on farms which had invested than on those that had not".
   (c) Test the prediction against reality by mounting a survey of farms. Analyse the results.
   (d) If the survey's findings agreed with the predictions, the hypothesis could be accepted.

1.10b No cause-and-effect relationship has been proved, just an association shown.

2.1 The man has a want—for his hunger to be satisfied. Goods which satisfy this want generate utility.
   Utility is generated by the ability of a commodity to satisfy a want. Demand is the desire for a commodity coupled with the ability to pay for what is desired.
   (i) Bread gives utility (it can satisfy the want) but we do not know if the man has the ability to pay for the loaf—demand is not established.

(ii) Bread gives utility but there is no demand because the man has no purchasing power.

(iii) Bread gives utility and there is demand—the man can afford to acquire the loaf.

(iv) Bread cannot satisfy the man's want—he is averse to it. Bread therefore generates no utility. Demand is therefore automatically absent.

(v) No utility or demand.

(vi) To a man who is fully satisfied a loaf of bread will give no additional satisfaction, i.e. it will generate no utility.

Demand cannot exist if there is no desire for a commodity.

2.2 The total and marginal utility schedules are:

Units of X	Total Utility	Marginal Utility
1	9	9
2	21	12
3	35	14
4	50	15
5	65	15
6	79	14
7	91	12
8	100	9
9	105	5
10	105	0

They have been plotted in the accompanying graphs.

2.3 The "Law of Diminishing Marginal Utility" states that "the utility of additional units of a commodity to any consumer decreases as the quantity of that commodity he is already consuming increases".

2.4 Yes, the law should really start "Once a consumer's stock of a commodity has reached a certain size, the utility of additional units ... etc.". The point at which diminishing marginal utility sets in is called the "Origin".

2.2 Total and Marginal Utilities

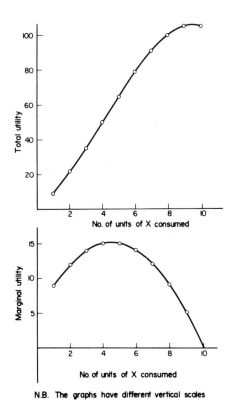

N.B. The graphs have different vertical scales

2.5 Car wheels to a motorist. Dining chairs which form a set? Another interesting example is noise and housewives. In a totally silent environment a woman alone at home all day can feel isolated and lonely. Some sound from neighbours or passing traffic can be a source of satisfaction if it relieves the feelings of being cut off.

More sound in an initially very quiet environment could give increasing marginal utility but soon diminishing marginal utility would set in. As the decibels built up, a level would be reached beyond which further increases would cause a reduction in satisfaction—sound would have become noise pollution and the marginal utility of sound would have fallen to negative quantities.

2.6 "Assuming that the Law of Diminishing Marginal Utility applies to both A and B, for maximum satisfaction this consumer should buy *more* of A and *less* of B. This adjustment of the purchasing pattern will make the ratio of Marginal Utilities closer to the ratio of prices."

2.7 (i) The indifference curves are drawn on the accompanying graph.
(ii) Marginal Rate of Substitution is the number of units of Y which have to be given up for an increase of one unit of X to maintain a constant level of satisfaction.
An indifference curve joins all combinations of amounts of two commodities (X and Y) which give a consumer the same level of satisfaction.

MRS	I_1 (X5.X10)	I_2 (X15.X20)	I_3 (X20.X25)
	2.0	1.2	1.2

(iii) Budget lines are shown on the graph for Q.2.7.

	Quantities purchased per week	
	X	Y
(a) Income £60 X = £2 each, Y = £2 each	10	20
(b) Income £88 X = £2 each, Y = £2 each	20	24
(c) Income £110 X = £2 each, Y = £2 each	25	30
(d) Income £110 X = £2 each, Y = £3 each	25	20
(e) Income £110 X = £2 each, Y = £6 each	25	10

2.8 From 16(iii), (c), (d), (e), the demand schedule for commodity Y is

Price of Y	Quantity Demanded per Week
£2	30
£3	20
£6	10

This demand schedule is plotted as a demand curve on the accompanying graph.

2.7   Indifference Curves and Budget Lines

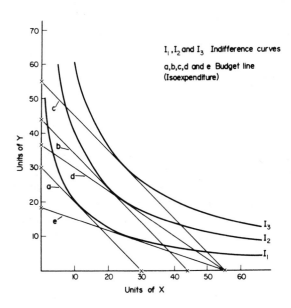

I_1, I_2 and I_3 Indifference curves

a,b,c,d and e Budget line
(Isoexpenditure)

2.8   Demand Curve for Commodity Y

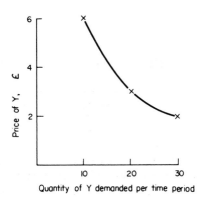

2.9 "A consumer is said to be in equilibrium when the Total Utility from his purchases is maximised. This also implies that the satisfaction he derives from his purchases is *maximised*. He allocates his spending, probably subconsciously, to this end by practising the Principle of *Equimarginal Returns*. The consumer's equilibrium is termed *stable* since any departure from this position caused by external forces will call into play forces which tend to restore the equilibrium position."

3.1 The total market demand for onions is found by adding together the demands of the three income groups at each price;

Price per Pound	Total Market Demand (thousand lb per week)
10	24
9	29
8	37
7	47
6	58
5	71
4	86

3.2 "If the price elasticity of demand of a good is equal to minus 3, then a 1 per cent fall will *raise* the quantity *demanded* by *3 per cent*. The total revenue of the seller of the good will *increase* and the total expenditure of the buyers of the good will *increase.*"

3.3 Income Elasticity of Demand of a good

$$= \frac{\% \text{ change in quantity (or expenditure) of the good demanded}}{\% \text{ change in consumer income}}$$

The percentage change in expenditure on housing is given (20%). The percentage change in income is $\frac{400}{2000} \times 100 = 20\%$

$$E_{DY} = \frac{20}{20} = 1$$

3.4 Engel's Law states that "the proportion of personal expenditure devoted to necessities decreases as income rises".

(b) and (c) are compatible with this law. Statement (a) is not because, in absolute terms, the same or more may be spent on essentials, although their relative proportion of the total spending will decline.

3.5 "The Elasticity of Supply with respect to product price is estimated as the percentage change in quantity supplied per time period *divided* by the percentage change in price. The Elasticity of Supply is generally *lower* in the short run than in the long run and this implies that the long-run supply curve is generally *less* steep than the short-run curve. In addition, the Elasticity of Supply will be *lower* if the producer has no alternative lines of production open to him; if the production cycle is short the Elasticity of Supply will be *greater* than if it is long."

3.6

a

(1) Change in consumer incomes

(2) Change in prices of complementary or competitive goods

(3) Change in tastes of consumers

(4) Change in population size

b

(1) Change in the prices of other goods the firms could produce.

(2) Change in the cost of factors of production.

(3) Change in price of joint products.

(4) Change in the state of technology.

(5) Change in the goals of firms.

3.7 Market for carrots; these answers are illustrated on the accompanying graph.

(i) Plotting the schedules produces a "scissors" graph.

(ii) The equilibrium price is 5p.

(iii) (a) At a market price of 6p demand would not be sufficient to clear the quantity of carrots supplied to the market. With time this surplus would build up until such time as the Government decided either to drop the price or to remove the surplus by physical destruction or the operation of a discriminating monopoly (including dumping abroad).

Answer 3.7

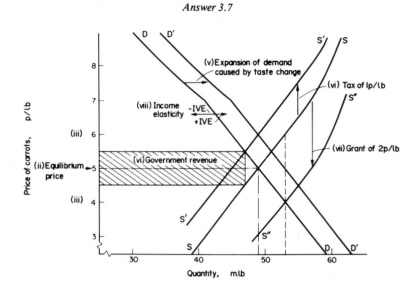

(b) A market price of 4p would cause demand at that price to exceed supply. Methods other than the price system would arise to distribute the carrots e.g. queuing, rationing, and a blackmarket might spring up.

(iv) (a) The Government will need to purchase 53m lb/wk.

(b) Its expenditure (price × quantity) will be £3.18m.

(c) The demand curve shows that consumers will demand this quantity only if the price is 4p/lb.

(d) The Government will receive from selling carrots (price × quantity) £2.12m.

(e) The Government's loss is expenditure minus revenue: £1.06m loss.

(v) If demand increases by 4m lb at all prices, the demand curve will move uniformly to the right by 4m lb. As demand has changed without any shift in the supply curve, the supply curve is cut at a higher price

New equilibrium price 5½ p
Quantity sold 51m lb

(vi) The conventional method of showing the effect of such tax is to shift vertically the supply curve by the extent of the tax. This assumes that producers pay the tax and add it to the price at which they sell. If producers were willing to supply 60m lb at 8p before the imposition of a tax of 1p/lb, then after the imposition of the tax the price which the market will need to pay to cause farmers still to produce 60m lb will be 8p + 1p = 9p. This is repeated at all levels of output. Thus a new supply curve is created above the original. This new supply curve cuts the original demand curve at 5½p, when 47m lb will be bought.

N.B. When the market price is 5½p, the Government will raise 1p × 47m lb = 47m p revenue from the tax. Producers will receive 4½p of the market price. This is a ½p less than they were receiving before the tax imposition. Consumers are paying ½p more than they were before the tax (5½p as opposed to 5p). The INCIDENCE of the tax in this case is said to fall equally on the producer and the consumer, as both suffer a similarly disadvantageous price change of ½p/lb.

(vii) Farmers given a grant of 2p/lb will be willing to supply to the market at 7p that quantity they were previously willing to supply at 9p and so on. A new market supply curve will be created vertically below the original one. This will cut the demand curve at a market price of 4p, when 53m lb will be bought.

(viii) (a) If the demand for carrots has a negative income elasticity of demand, an increase in consumers' incomes will cause demand for carrots to contract, i.e. the demand curve will shift to the left. From the scissors graph it can be seen that such a shift will cause prices to fall.

(b) The argument is the reverse of (a).

3.8   (i) Slope of this straight line is − 1.

      (ii) Price elasticity of demand

$$= \frac{\text{percentage change in quantity demanded}}{\text{percentage change in price}}$$

The prices are given. The quantities demanded at these prices must be read from the graph.

(a) *Example*

Price 9.2 units quantity demanded 0.8 units
Price 8.8 units quantity demanded 1.2 units.

$$E_{Dp} \quad \frac{\dfrac{.4}{8} \times 100}{\dfrac{-.4}{9.2} \times 100} = \frac{50\%}{-4.4\%} = -11.4$$

(b) -2.6

(c) -0.45

From a, b, and c we can say:
"The price elasticity of a straight-line demand curve is *not* constant along its whole length".

(iii) Using the formula given, which measures the elasticity at single points, the following results are obtained:

(a) at price 2 units E_{Dp} is -0.25
(b) at price 6 units E_{Dp} is -1.5
(c) at price 8 units E_{Dp} is -4.0

From the formula

$$E_{Dp} = \frac{1}{\text{Slope at price X}} \times \frac{\text{Price X}}{\text{Quantity demanded at price X}}$$

it can be seen that, where two curves cross, i.e. price and quantity demanded are the same for both curves, the price elasticity of demand of each curve is inversely proportional to the slope of the curve. The curve with the steeper slope has the elasticity coefficient nearer to zero, i.e. is the less elastic. Hence the second sentence of the statement should read thus: "At the point of intersection the steeper curve possesses the lower price elasticity of demand, and is therefore said to be the *less* elastic of the two".

4.1 In a perfectly competitive industry:

(a) there are *many* producers
(b) entry to the industry is *free*
(c) each producer *cannot* influence market price.

In a complete monopoly:

(a) there is *one* producer

(b) entry to the industry is *restricted*

(c) the producer has *considerable* influence on price.

To a *producer* in a perfectly competitive industry:

(a) Marginal Revenue *is* constant with varying levels of output

(b) Marginal Revenue *is* identical with Average Revenue

(c) the demand curve which he faces is infinitely *elastic*.

To a *producer* in an imperfectly competitive industry:

(a) Marginal Revenue *is not* constant with varying levels of output

(b) Marginal Revenue is normally *less* than Average Revenue

(c) the demand curve for his product is normally *down-sloping*.

A perfectly competitive *industry* faces a demand curve which is *less than infinitely* elastic.

The Marginal Revenue of a perfectly competitive *industry* normally *declines* with increasing output.

4.2 Schedules for the sale of rubber boots.

Price per pair £	Quantity Demanded per week	Total Revenue £	Marginal Revenue £
2.75	1	2.75	2.75
2.45	2	4.90	2.15
2.20	3	6.60	1.70
1.90	4	7.60	1.00
1.65	5	8.25	0.65
1.45	6	8.70	0.45
1.20	7	8.40	− 0.30
1.00	8	8.00	− 0.40
0.80	9	7.20	− 0.80
0.60	10	6.00	− 1.20

(a) Price per pair which the shopkeeper receives is identical with his Average Revenue for pairs of boots. Hence plotting Average Revenue and Marginal Revenue involves the plotting of the first and fourth columns above.

(b) The price of boots when Marginal Revenue is £1.00 is £1.9 and the quantity sold at this price is 4 pairs. This answer can be obtained either directly from the table, or from the graph. The Marginal Cost of boots is constant in this case at £1, so that the shopkeeper's Marginal Cost curve is a horizontal straight line at £1 on the vertical axis. This line will cut the MR curve, and the quantity sold and average revenue (price) can be read.

N.B. the AR curve is the same as the demand curve faced by the producer (shopkeeper). It shows the relationship between price and quantity sold. The slope of the demand curve at price £1.9 is about -0.22. (The exact measurement of slope is unimportant.)

Using the formula to estimate price elasticity of demand

$$E_{Dp} = \frac{1}{\text{Slope at price X}} \times \frac{\text{Price X}}{\text{Quantity demanded}} = \frac{1}{-.22} \times \frac{1.9}{4} = -2.2$$

Hence the price elasticity of demand at this most profitable level of sales is greater than $(-)$ one. This agrees with the prediction that the most profitable level of output for a monopolist occurs when he has restricted his output where the demand for his produce is elastic (i.e. price elasticity of demand coefficient greater than one). His MR at this point will be positive.

If the agricultural *industry* is like a monopolist in the supply of food, and if the price-elasticity of demand for its products is between -1 and 0 can you suggest what it could do to increase its profit?

4.3 (a) Initial distribution of national income by spenders.

National Income	Agriculture		Other industries	
	receives in exchange for its products	% of national resources receiving this	receive in exchange for their product	% of national resources receiving this
£100m	£50m	50%	£50m	50%

Income elasticity of demand $= \dfrac{\text{\% change of demand for a commodity}}{\text{\% change in consumer income}}$

A 10% rise in national income is synonymous with a 10% rise in the income of consumers making up the nation. This will cause an increase in demand, the size of the increase for each product being reflected in the particular income elasticity of demand for that product.

Percentage increase in demand (by re-arranging the formula above).

> *Agriculture:*
> % change in demand = 0.4 × 10% = 4%
> *Other industries:*
> % change in demand = 1.6 × 10% = 16%

Expansion in expenditure:

> *Agriculture:* a 4% increase on £50m is £2m.
> *Other industries:* A 16% increase on £50m is £8m.

The final distribution of national income by spenders is:

National Income	Agriculture	Other industries
£110m	£52m	£58m

After the increase in national income the half of the nation's resources used in agriculture is attracting a lower share of the national expenditure than the half in the other industries. In the initial state the resources inside and outside agriculture were rewarded to the same extent, i.e. there was parity of reward between the two sectors of this simplified economy, whereas after the rise in national income this parity is destroyed, with the 50% of national resources used in "other industries" attracting more than 50% of total expenditure. This is reflected in higher wages, interest rates, rents and profits in "other industries". One would expect resources (land, labour, capital, management) to be attracted from agriculture by these higher returns. However, resources used in agriculture are notoriously slow to move to other forms of production, resulting in the rewards in other industries being persistently higher and leading to demands for income support to agriculture.

The statement in 4.3 should read:

"After the rise in national income the half of the nation's resources in agriculture is earning *less* than half in the other industries. Resources should be transferred *from* agriculture *to* the other industries if the returns to the nation's resources are to be equated in each sector."

4.4 "If a firm in perfect competition is attempting to maximise profits it should produce that level of output where the addition to total costs caused by producing the last unit of output (called *Marginal* Cost of production) just equals *Marginal* Revenue. In perfect competition this will also *equal* the price of the product. A monopolist *will* achieve maximum profits by pursuing the same policy and in his case the Marginal Revenue at his optimum level of output will be *less than* the price which he charges for his product."

4.5 Compared with a non-discriminating monopolist, a monopolist engaged in price discrimination will generally have a *greater* revenue. This is because the discriminating monopolist equates *marginal* revenue for his product in each of the markets between which he discriminates.

Answer 4.6

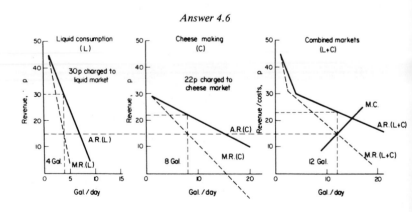

4.6 (a) Two or more markets for the monopolist's product which have different demand characteristics.

(b) The ability to prevent "leaks" or "seepage" between the markets by legal, time, distance or other constraints.

Quantity sold (Gal/day)	Liquid consumption		Cheese-making	
	Total Revenue £	Marginal Revenue £	Total Revenue £	Marginal Revenue £
1	0.45	0.45	0.29	0.29
2	0.80	0.35	0.56	0.27
3	1.05	0.25	0.81	0.25
4	1.20	0.15	1.04	0.23
5	1.25	0.05	1.25	0.21
6	1.20	− 0.05	1.44	0.19
7	1.05	− 0.15	1.61	0.17
8	0.80	− 0.25	1.76	0.15
9	0.45	− 0.35	1.89	0.13
10			2.00	0.11
11			2.09	0.09
12			2.16	0.07
13			2.21	0.05
14			2.24	0.03
15			2.25	0.01
16			2.24	− 0.01
17			2.21	− 0.03
18			2.16	− 0.05
19			2.09	− 0.07
20			2.00	− 0.09

(c) 30p (see accompanying graph)
(d) 4 gal
(e) 22p
(f) 8 gal.

5.1 The production function is drawn on the graph to accompany this answer.

(i) The point of inflection of the production function *Marginal Product* is maximal.

(ii) At the point where a line from the origin is tangential to the production function *Marginal Product* and *Average Product* are equal.

Answer 5.1

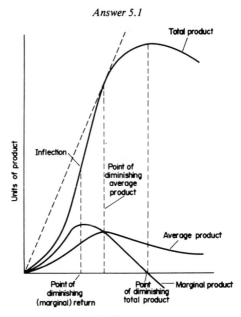

Answer 5.2

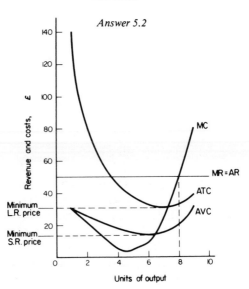

(iii) When Total Product is maximal *Marginal Product* is zero.

(iv) When Total Product is declining with increasing quantities of input *Marginal Product* is negative and *Average Product* is positive and *declining*.

(v) The point where Marginal Product is maximal is also called the *Point of Diminishing (Marginal) Returns.*

(vi) Where Total Product has been reduced to zero by using a very high level of input. Average Product is also zero. An example might be where extremely high levels of nitrogen fertiliser are used on cereals. Lodging of the crop, disease and killing by excessive nitrogen may reduce the harvestable crop to nothing.

5.2 The completed table is

Output (units)	Total revenue	Marginal revenue	Total costs	Fixed costs	Variable costs	Marginal cost	Average total cost	Average variable cost
	£	£	£	£	£	£	£	£
0	—	—	110	110	0	—	—	—
1	50	50	140	110	30	30	140	30
2	100	50	162	110	52	22	81	26
3	150	50	175	110	65	13	58.3	21.7
4	200	50	180	110	70	5	45	17.5
5	250	50	185	110	75	5	37	15
6	300	50	194	110	84	9	32.3	14
7	350	50	219	110	109	25	31.3	15.6
8	400	50	269	110	159	50	33.6	19.9
9	450	50	349	110	239	80	38.8	26.5

(a) The firm is producing under perfect competition as its MR does not fall with increasing output.

(b) Fixed costs are those which it has to bear whether or not it produces anything, and so are its costs when output is zero. Total costs when output is zero are all fixed costs, and are £110.

(c) The table is plotted on the graph accompanying this answer.

(d) A firm is in equilibrium when *Marginal Cost* equals *Marginal Revenue* (Marginal Cost is the increase in Total Cost caused by the last unit of output and Marginal Revenue is the increase in Total Revenue caused by the last unit of output.)

(e) 8 units of output.

(f) (i) Total Revenue (£400) minus Total Costs (£269) = £131.

    (ii) Average Revenue (£50) minus Average Total Costs (£33.6) multiplied by the number of units of output (8).

$$(£50 - £32.6) \times 8 = £131.2$$

This is the maximum profit that can be earned by the firm. Try subtracting Total Costs from Total Revenue at other levels of output. Lower profits will always be found.

(g) (i) The price of an article is synonymous with its Average Revenue. In the long run a firm willl need to cover its fixed costs e.g. capital depreciation, rent and rates, regular labour force, interest charges on borrowed capital, and if the average price it receives for its product is not sufficient to at least cover the fixed and variable costs, it will eventually go out of production. From the graph it may be seen that if the price of its product is about £31,* then the firm will just cover its fixed plus variable costs if it produces 7 units of output. If the price were below £31, at no level of output would its Average Total Costs be covered. At higher prices the firm will have a range of outputs where price (Average Revenue) exceeds Average Total Costs, and within this range is situated the most profitable level of output. At this most profitable level, Average Total Cost will not be at its lowest point, but on the rise.

    Minimum long run price—£31

    Quantity produced—7 units.

5.3 (a) The isoquants are given in the graph accompanying this exercise.

(b) An iso-cost line must be constructed. This connects the quantities of oats and barley which can be bought for the same expenditure. If the prices of the cereals are equal at £30/ton, then say £120 will buy 4 tons of each, or any combination of the two lying along the iso-cost line connecting 4 tons of barley to

---

* In this and the other examples in this section where values have to be read from graphs some tolerance must be allowed because of the slightly differing ways in which individuals draw curves.

Answer 5.3

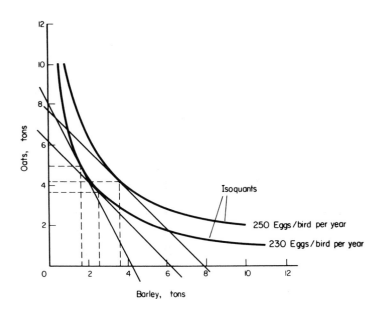

4 tons of oats. Iso-cost lines further away from the origin represent higher levels of cost, but the slopes will not alter as long as the relative prices of oats and barley are constant. Hence, once one iso-cost line has been drawn it may be slid parallel until it is tangential to an isoquant, and the optimum combination of the two cereals read off. This is the combination which produces the quantity of eggs represented by the isoquant at the lowest cost. Other combinations *may* be used, but these will correspond with iso-cost lines which are further from the origin (i.e. more costly) and which *cut* the isoquant.

Optimum quantity to feed to achieve 230 eggs/bird/year
    Oats 3.6 tons
    Barley 2.6 tons

(c) Relative proportions
    Oats/Barley = 3.6/26 = 1/0.7

(d) Oats   3.6 × £30 = £108
     Barley   2.6 × £30 = £ 78
     Total cost   £186

(e) Slide the original iso-cost line until it is tangential to the second isoquant.

         Oats 4.2 tons
         Barley 3.6 tons

The proportions have changed in this higher production diet as well as the absolute quantities. Oats/Barley is now 1/0.86. This change is caused by difference between the isoquants. It is unlikely that the proportions will remain the same as production is increased.

(f) A new iso-cost line must be constructed. The new optimum is
         Oats 5 tons
         Barley 1.6 tons

(g) As might be expected, a fall in the price of oats has caused more of them to be used. (Compare the quantities in (f) with those in (b) which produce the same number of eggs.) If barley is still at £30/ton, then the total cost of the diet will have fallen.

         Oats   5 ton × £15/ton = £75
         Barley 1.6 ton × £30/ton = £48
                             £123 (as opposed
                                       to £186)

5.4 (a) The production possibility curve is plotted on the graph which accompanies this exercise.

(b) (i) The MRS of products at one level of resource use is the change in the number of units of product on the vertical axis which occurs when production of the product on the horizontal axis is increased by one unit. With products which compete for the firm's resources an increase in production of one product is accompanied by a fall in the other, and the MRS is negative (i.e. a negative change divided by a positive change). Over the range quoted in this part of the question hay and wheat are complementary, an increase in hay production also producing an increase in wheat production.

$$\text{MRS} = \frac{\text{Change in wheat production}}{\text{Change in hay production}} = \frac{+50 \text{ tons}}{+50 \text{ tons}} = +1$$

(ii) Over this range the products are competitive.

$$\text{MRS} = \frac{\text{Change in wheat production}}{\text{Change in hay production}} = \frac{-60}{+50} = -1.2$$

(c) To find the most profitable combination of products an iso-revenue line must be constructed. It links the quantities of hay and wheat which bring in the same revenue. All combinations of hay and wheat lying on this line will also bring in that same revenue. Iso-revenue lines further from the origin represent higher (and therefore more desirable) levels of revenue. The optimum combination of products is where the production possibility curve touches an iso-revenue line furthest from the origin.

If wheat and hay are the same price per ton, then an iso-revenue line would connect (say) 200 tons hay with 200 tons wheat, as both these quantities would bring in the same revenue. This line may be slid parallel away from the origin to represent higher iso-revenue lines. Where the production possibility curve just touches an iso-revenue line is the optimum combination of products. Note that it is the *relative* prices of the products which determines the slope of the iso-revenue line and hence the optimum combination.

  (i) Wheat 260 tons
     Hay   100 tons

 (ii) A new iso-revenue line is required from (say) 100 tons wheat to 200 tons of hay—both of these will bring in the same revenue. Proceed as above.
     Wheat 260 tons
     Hay   100 tons

(iii) As (ii) above
     Wheat 260 tons
     Hay   100 tons

(d) If hay produces no revenue than an infinite amount of it will be necessary to produce the same revenue as, say, 200 tons wheat. In this case, therefore, the iso-revenue line is horizontal and can be slid up and down to find the optimum combination of the two products, which will be 260 tons of wheat and 100 tons of hay. Even when hay produces no revenue it will be desirable to produce 100 tons, as this increases wheat production.

For convenience, this example has been constructed with a sharp angle to the production possibility curve; this produces identical answers for the various price levels quoted. In practice a smoother curve is likely and the optimum combinations are likely to be more sensitive to changes in relative prices.

5.5   Option c is the best description of a Fixed Cost.

5.6   "If Marginal Cost is rising with increasing output, Average Total Cost *may be rising or falling.* Average Fixed Cost will be falling. The minimum price which the entrepreneur will be able to accept in the short run will correspond to the lowest point on the *Average Variable Cost* curve, but in the long run price must be at least the lowest point on the *Average Total Cost* curve. The rising Marginal Cost curve cuts both the *Average Variable Cost* curve and the *Average Total Cost* curves at their lowest points. To make maximum profits an entrepreneur will select that level at which *Marginal Revenue* equates with *Marginal Cost.*"

5.7   Refer to the text to check your answers.

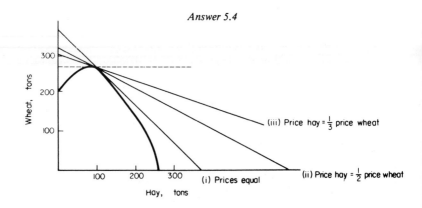

Answer 5.4

6.1 The demand for a factor of production is a derived demand as the factor is not wanted for its own sake but for the contribution it makes to the production of goods which *are* wanted for the satisfaction of wants. The demand for factors is derived from the demand for these want-satisfying goods. Anything which affects the demand for these goods will affect the demand for the factors producing them. The products of one firm (e.g. tractors) may be factors of production of other firms (e.g. farmers). The demand for steel for tractor making is derived from the demand for tractors which in turn is derived from the demand for agricultural products.

6.2 Five different goods from which timber derives its demand.
(i) furniture, (ii) houses, (iii) boats, (iv) garden sheds, (v) magazines (via paper).

6.3 The price elasticity of demand for a factor depends on:
(i) the price elasticity of demand for the product,
(ii) the cost of the factor relative to total costs of production,
(iii) the presence (or absence) of good substitutes for the factor.
   (a) Rises in the price of glass will cause prices of glasshouses to rise. A price-elastic demand for glasshouses will result in a considerable fall-off in the quantity of glasshouses sold, which will be reflected as a fall in the quantity of glass demanded. The more elastic the demand for glasshouses, the more elastic will be the demand for glass.
   (b) If the cost of water represents only a very small part of the total costs of bread-making, the price of water can vary a lot before it significantly alters the price of bread and the quantity of bread demanded and subsequently is reflected back in a reduced demand for water. The demand curve for water is hence very steep and the demand price-inelastic.
   (c) The presence of good substitutes means that producers will switch between factors as the prices of competing fertiliser brands change. Small changes in the price of one brand cause large changes in the quantity demanded— demand is price-elastic. For fertiliser in general, however, there is no good substitute and the general level of price can fluctuate widely with relatively small changes occurring in the quantity demanded.

6.4   A farmer, given a certain quantity of farm labour, must so allocate it among his various enterprises that the *last* man-hour used in each enterprise yields *the same* value of output. This will, of course, *not mean* that the same amount of labour must be devoted to each enterprise.

This is an example of the Principle of Equimarginal Returns. If labour were not allocated in this way the farmer would gain by moving labour from where it was earning a low Marginal Revenue Product to where it was earning a higher Marginal Revenue Product.

6.5   To use the Principle of Equimarginal Returns one must derive the Marginal Products of each day of cultivation on each field from the Total Product figures (yield) given. The Marginal Product of, say, the 3rd day of cultivation is the yield from 3 days' work minus the yield from 2 days' work and so on.

Marginal Product of Labour (Tons of Potatoes)

Days of cultivation	Field A Tons of potatoes	Field B Tons of potatoes	Field C Tons of potatoes
1	15	20	10
2	10	10	7
3	8	9	6
4	4	6	1

To derive the best use of his 8 days cultivation time he must obviously use all 8 days. The Table readily shows what the time allocation would be if he had only 5 days, as the Marginal Product of a day's cultivation is 10 tons of potatoes on each field if he spends 2 on A, 2 on B and 1 on C. The best use of the 6th day would be on B, as its Marginal Product is greatest (9 tons). The 7th day would be best spent on giving A a third day of cultivation, (Marginal Product 8 tons) and the 8th day best spent on C (Marginal Product 7 tons).

The best allocation of 8 days of cultivation is thus

Field A   3 days
Field B   3 days
Field C   2 days

The total yield is 96 tons from A, 79 tons from B and 82 tons from C, totalling 257 tons. Any other time allocation will give a lower overall yield.

6.6   (i) Farmer A is geographically *mobile*, as he is willing to move around the country, but occupationally *immobile* as he is unwilling to leave sheep farming.

  (ii) His labour is specific—he cannot be switched easily to other industries (or even within farming it appears).

  (iii) Specific labour is often geographically mobile—e.g. brain drain to USA, opera singers jet-hop around the world's opera houses.

  (iv) Non-specific labour is usually the more occupationally mobile—e.g. a labourer can work in many industries.

  (v) The more specific are   (i)   road roller
                        (ii)   cow-kennels
                        (iii) moorland
                        (iv) old farmer

They are specific because they cannot be easily switched to uses other than those in which they are normally found.

6.7 Economic rent: (i) Economic rent is the surplus received by a factor above its tranfer earnings.

  (ii) A £100,000/yr. actor who has a best-paid alternative employment yielding £20,000/yr. has earnings which consist of £20,000/yr. transfer earnings, and £80,000/yr. economic rent.

  (iii) The actor receives economic rent in the payment for his services because (a) his services are economically scarce, i.e. they are limited in supply relative to the desire for them, and (b) his services are inelastic in supply. If demand for his services increases the price of his services will rise, and an even larger proportion than before will consist of economic rent.

  (iv) Other people will be attracted into the theatrical profession in the long run by high economic rents if these contain an element of pure profit. Incomes will tend to be pushed down, and so economic rents in the profession as a whole will decline. However, for individual good actors there are no perfect substitutes, and as their services are always scarce (simply because there is demand for their unique talents) they will always earn above their transfer earnings.

6.8   (i) By definition, risks are possibilities of future loss which can be formally insured against. Entrepreneurs thus can remove risk by means of insurance policies.

(ii) Two sources of risk in agriculture, both insurable, are loss by fire and loss by personal accident.

(iii) Sources of uncertainty.

    (a) Changes in price and/or costs between when production is initiated and when its products are ready for sale. The longer the production period the greater the uncertainty, *ceteris paribus*.

    (b) Failure of physical production (yields etc.) caused by weather, disease etc.

    (c) Changes in Government policy.

    (d) Changes in the other sectors with which farmers trade directly or indirectly. The invention of new man-made fibres may reduce the demand for wool. Vegetable, milk and cream substitutes may cut back the demand for the natural products, etc.

    (e) New production processes may be developed in agriculture which increase production, lower price and place those farmers not adopting the new techniques in untenable positions e.g. the widespread use of yard and parlour systems; c.f. cow-house milk production, bulk handling of grain made bigger combines uneconomic to operate.

    (f) The personal circumstances if the farmer and family have a great influence on the farm business, e.g. death of a father may necessitate sale of part of the farm to pay capital taxes.

(iv) Reduction of uncertainty is effected by:

    (a) choice of enterprises which are relatively safe;

    (b) diversification within the farm business i.e. having more than one enterprise on the farm;

    (c) production flexibiity (i) cost flexibility: e.g. the choice of a system of production with low fixed costs enables an entrepreneur to scrap an enterprise more easily; (ii) product flexibility: processes which can switch products are less subject to uncertainty; (iii) time flexibility: enterprises with short production cycles tend to be safer than those with long cycles, *ceteris paribus;*

(d) informal insurances: e.g. potato blight can be controlled by preventative spraying;

(e) contracting and vertical integration: contracts both for inputs and products can limit cost and price uncertainty, and ensure supplies and outlets for products;

(f) the use of economic and statistical data and research: any producer who is not aware of the economic characteristics of his product—price cycles, long term trends etc.—is in a worse position to react to changes than his informed fellow-producers;

(g) hedging: dealing in futures.

6.9 (i) When prices and costs are stable an industry in perfect competition will so arrange itself that its entrepreneurs are earning *normal* profits. If the beef-producing industry is in such an equilibrium, and the demand for beef suddenly increases because of an increase in consumers' incomes, in the short run beef producers will earn surplus profits. New entrepreneurs will be attracted into the industry and force *down surplus* profits until a level is restored where entrepreneurs are again earning *normal* profits.

(ii) "Normal profit is the payment necessary to keep an entrepreneur in a particular line of production."

(iii) Normal profit is a return to entrepreneurship and so interest on capital will have already been removed.

(iv) He will leave the industry if he persistently earns less than normal profits.

(v) Entrepreneurs will require a high normal profit in industries where the possibility of a loss is high. The factors of production put into making a record which flops are lost. Differences in uncertainty determine differences in normal profits between industries and enterprises within industries.

7.1 Any of them might be reasons why the Government wished to involve itself in the economy.

7.2 Only (i) is an external diseconomy. Both (iii) and (v) are external *economies* (or benefits).

7.3 From society's viewpoint the optimum amount of a commodity to produce is when marginal *social* cost is equated with marginal *social* benefit.

7.4 (i) Private cost of bricks is £x per ton.
(ii) Total external diseconomy is £z.
(iii) Social cost of bricks per ton is £(x + z /y).

7.5 Ways in which an external diseconomy may be taken into consideration in the production and consumption decisions of society include banning (drugs), prescribing maximum quantities (pollution of water), taxing (heavy lorries), advice (agricultural smells), education (use of degradable plastics), compensation (compulsory for damage caused to health of workers), taking into public ownership (to protect wildlife sanctuaries from other users).

7.6 Ways in which an external economy may be taken into consideration in the production and consumption decisions of society include compulsion (taxes to finance defence and police), levies (to support training), financial incentives (for environmental conservation), advice and education (to farmers on the heritage worth of their buildings).

7.7 Pure Public Goods are non-excludable and non-rival.

7.8 All can be considered as Public Goods except public buses. Only on buses can the seat be something which one person can prevent another from using and the more seats one passenger occupies the less are available for other people.

7.9	Publicly provided	Publicly funded
(i) NHS hospital care	yes	yes
(ii) State education	yes	yes
(iii) NHS dental care	no	yes
(iv) British Rail (making profits)	yes	no
(v) Royal Mail (making profits)	yes	no

Dentists are private practitioners who are paid by the State for the work that they do, though this is changing and a greater proportion of the cost is being borne by patients. Family doctors (General Practitioners) are also private operators but funded publicly. Until recently many of the utility industries (gas, electricity etc.) were publicly operated but consumers paid for their goods and services in much the same way as they paid for petrol or other private goods.

8.1 The principal injections into the circular flow are investment, Government spending and exports; the principal withdrawals are saving, taxation and imports.
   (a) Withdrawal
   (b) Injection
   (c) Injection
   (d) Withdrawal
   (e) Injection
   (f) Neither—the gift is simply a transfer of spending power. If, however, he saves it, a withdrawal will occur
   (g) Neither

8.2 The Marginal Propensity to Withdraw is the *proportion* of an increase in *total* income which is withdrawn by saving, taxation, purchase of imports etc.

8.3 The relationship between the Marginal Propensity to Save (MPS) and the Multiplier is thus:

$$\text{Multiplier} = \frac{1}{\text{MPS}}$$

$$\text{In this case} = \frac{1}{0.25} = 4$$

This means that the effect of £1 extra spending in the economy by investment or any other form of injection will cause total spending to rise by £4.

8.4 National Income will be in equilibrium when the intended level of injections is *equal to* the intended level of withdrawals. This equilibrium level *need not* correspond to one at which all the nation's productive resources are fully employed.

8.5 (a) We are given some important pieces of information about the economy illustrated in the diagram. The first is that the National Income is in equilibrium; this means that Savings (S) and Investment (I) must be equal because injections and withdrawals must be the same (other forms of injections and withdrawals are excluded in this simplified economy). If consumption spending (C) is £600,000 out of a National Income of £1m, then £400,000 must be saved (S). Because National Income is in equilibrium, this means that I also is £400,000.

(b) Of the £1m National Income, £600,000 is spent on consumption i.e. 0.6 of the total. Average Propensity to Consume is therefore 0.6. As the question states that APC = MPC, MPC is therefore 0.6.

(c) With a National Income of £1m and an MPC of 0.8 where MPC = APC, £800,000 will be spent on consumption and £200,000 saved. An equilibrium in National Income will mean that I will also need to be £200,000.

(d) For equilibrium S must equal I i.e. S must be £400,000. What level of Income will give this level of S? If MPC is 0.8, MPS must be 0.2, i.e. 0.2 of each £1 of income will be saved. The level of National Income (Y) to give £400,000 of saving is:

$$Y = \frac{£400,000}{0.2} = £2m$$

Consumption spending (C) will be $Y - S = £2m - £400,000 = £1.6m$.

8.6 (a) True—the CC line rises as National Income (Y) increases.

(b) True—at levels of Y below OL.

(c) False—when consumption (C) is greater than (Y) the difference must be made up by drawing on reserves, i.e. dis-saving or negative saving.

(d) False—the proportion falls as income rises. This proportion is also called the Average Propensity to Consume (APC)

$$\text{APC of any level of } Y = \frac{\text{Total Consumption}}{\text{Total Income}} = \frac{C}{Y}$$

This falling APC is also shown by the slope of the CC line being less than the 45° line which illustrates what would be the situation if all income, no more and no less, was all spent on consumption whatever the level of income might be.

(e) False—in this case the CC line is straight. If you were to measure how much consumption increased with successive small rises in income you would find that the same proportions of the extra income was always consumed and saved, i.e. MPC and MPS are constant

Note that although MPC is constant, APC is falling (see (d) above) because the CC line does not pass through the origin. If it did, APC would be constant and equal MPC. This is why in Exercise 7.4 you were told that MPC = APC; without that information you could not have deduced a level of APC from MPC.

(f) True
(g) True

8.7 (a) Raise—in that relaxing H.P. controls encourages consumer spending; an enlarged C will raise all the other curves.
(b) Lower—C will contract and maybe I too, taking the other curves with it.
(c) Raise—the G component of C + I + G will expand.
(d) Lower—C will contract.
(e) Lower—C and I will both contract because of the higher cost of personal and business loans.
(f) Raise—for the opposite reason to (e).
(g) Raise—C expands.

8.8 (a) The Deflationary Gap is the shortfall in aggregate demand, measured at the full employment level of National Income by which demand would need to be expanded to achieve full employment; RS in the diagram.

(b) Conversely to (a) the Inflationary Gap is QR. It is the amount by which aggregate demand should be reduced to produce an equilibrium National Income at the full employment level.

(c) Prices will rise due to demand exceeding output at the initially prevailing prices.

8.9 The most promising alternative is (b) as it will lower the level of aggregate demand.

(a) will increase aggregate demand and the Inflationary Gap;

(c) will also increase aggregate demand; some of the tax payment will be financed by reducing saving so that the amount of *consumer* expenditure curtailed by the taxes will be *less* than the extra *Government* expenditure;

(d) poor people have a lower propensity to save than the rich, so the transfer will cause overall savings to fall and consumption spending to rise, raising aggregate demand.

8.10 $M \times V = P \times T$

8.11 Demand-pull inflation and Cost-push inflation.

8.12 "In the Phillips Curve the *annual percentage change in the level of prices* is shown as being *inversely related to* the percentage level of unemployment, so that the higher the level of unemployment, the *lower* the level of inflation."

9.1 The commodity foregone—5 tons of wheat.

9.2 A country is said to possess comparative advantage *vis-à-vis* another country in the production of a good "X" if the expansion of production of good "X" by one unit involves foregoing the *production* of a *lesser* quantity of other goods in the first country than in the second.

9.3 No. Trade depends on comparative advantage, not on the absolute levels of efficiency. Other countries may have a comparative advantage over the UK in food production despite the high absolute efficiency of UK agriculture.

9.4 (a) Maxitry has the comparative advantage in peach production; its opportunity cost of expansion of peach production is lower in terms of coconuts foregone than Minitry's.

(b) Minitry.

(c) Minitry, since it would expand coconut production and so have a surplus for export in exchange for peaches.

(d) Both.

9.5 "Specialisation and trade should be in such a direction that Country A exports wheat and Country B exports maize."

This arrangement follows as A has the comparative advantage in wheat production. One extra ton involves the loss of 1 ton of maize, whereas in B it involves the loss of $6/4 = 1.5$ tons of maize.

9.6 Under perfect competition the relative prices of goods indicate the relative quantities of resources going into their production. For example, in Country C, 1 lb of beef takes 3 times the resources of 1 lb of butter. Every extra 1 lb of beef produced then would involve the loss of 3 lb of butter. In Country D an extra 1 lb of beef would involve the loss of only 2 lb of butter. Country D thus has a comparative advantage in beef production, since it has the lower opportunity cost. It follows that the answers are thus:

(a) Country D

(b) Country C

9.7 Yes, we can predict the relative prices of butter and beef, but with the information given we cannot predict a precise figure, only a range within which it will lie. It will be somewhere between the two original internal price ratios i.e. between 1 : 3 and 1 : 2.

9.8 (1) To protect infant industries.

(2) For defence purposes.

(3) To prevent dumping.

(4) To allow time for factors of production to be reallocated.

(5) For the protection of society against undesirable imports (drugs, illegal firearms etc.).

9.9 "The 'Terms of Trade' are said to 'improve' when the *index of export prices* rises relative to the *index of import prices.* Such an 'improvement' will have an effect on the Balance of Payments which *can be either beneficial or deleterious depending on circumstances.* "

9.10   a, b, c, d, f, h.

9.11   a, d.

9.12   b, c, f, h.

9.13   e (g is usually treated as a Monetary Movement).

9.14   b, c, e. Each of these increase the demand for foreign currency and increase the supply of £'s.

10.1   The correct order of the policy process is:
policy effect (4); policy implementation (3);
problem formulation (1); policy formulation (2).

10.2   The meanings of the terms are:
(a) evaluation; judging the outcome in relation to objectives and resources used;
(b) monitoring; collecting data during a policy programme for use in evaluation;
(c) base-line study; a look at the situation before a programme starts;
(d) side-effects; effects felt outside the programme.

10.3   Types of policy intervention that might be considered to reduce river contamination by silage effluent would include the following: education of farmers of the damage caused by effluent; offers of free Government advice on effluent disposal; grants to assist with the installation of control equipment; laws on maximum permitted effluent emissions, with fines.

10.4   When agriculture is described as being used in an instrumental role, the policy aim is one not immediately concerned with agricultural production, for example in creating rural jobs, or in producing a cared-for countryside.

10.5   Reasons why a government may wish to have an agricultural policy include the promotion of food security, economic growth, rural jobs and a vital rural economy, farm incomes, labour mobility, balance of payments, environmental appearance, and others.

10.6 The Government's intervention buying creates a new demand curve, as seen from the suppliers' viewpoint, which is infinitely elastic at £100; refer back to Fig. 10.5 in this chapter. Suppliers will expand their output in response to the higher price.

(a) The Government will buy 3m tonnes (the difference between the 7m tonnes produced and the 4m tonnes which other buyers are willing to take from the market, which is less than they took at the previous lower market price).

(b) It will spend £300m.

(c) It will lose £150m (what it spends minus its revenue from selling on the world market).

(d) The output of UK wheat expands from 5m to 7m tonnes.

10.7 Continuing the same example, what would you expect to happen to the following? Indicate the direction of change (*up/ down* or *unchanged*).

(a) The output of other cereals would fall as wheat was substituted for them.

(b) The income of wheat producers would rise.

(c) The demand curve of the users of wheat remains unchanged.

(d) The amount of wheat that is bought by users falls as they retreat up their demand curve.

(e) The profits of firms who use wheat will fall—the factor of production has become more expensive. The extent of the fall will depend on the ability to use alternatives and the elasticity of demand of the products.

(f) The prices of the consumer products will rise.

(g) The price of land which can grow wheat will be bid up—it reflects an increased derived demand.

(h) the demand for chemicals specific to wheat-growing will rise.

(i)   the price of these chemicals will be unchanged if their supply is completely elastic, but rise if it is less than completely elastic.

10.8   (i) Before the tax is imposed:
(a) 2m tonnes are produced in the UK.

(b) 6m tonnes are consumed.

(c) 4m tonnes are imported.

(d) £400m are spent on imports.

(ii) After the tax is imposed:
(e) 4m tonnes are produced in the UK.

(f) 5m tonnes are consumed.

(g) 1m tonnes are imported.

(h) £100m are spent on imports by the nation (though consumers have to pay £120m because they face a higher price than the world price).

(i)   £20m are raised in taxation. (This accounts for the difference in (h) above.)

10.9   (a) The highest price that can be arranged for farmers by taxing imports is the price when imports are taxed so highly that none come in. Price is then determined by domestic supply and demand. This price is £125.

(b) The minimum level of tax (per tonne) which brings this about is £25.

(c) None is imported, so the revenue is nil.

10.10 (a) The deficiency payment is £20 per tonne.

(b) 2m tonnes are suppied by UK producers before the introduction of the payment.

(c) When the payment is introduced they supply 4m tonnes; the Government expenditure is therefore £80m.

(d) Because the price to consumers is not altered, there is no change in the quantity demanded from the market.

(e) The amount imported falls by 2m tonnes.

Comparison between the answers to 10.9 and 10.10, which use the same basic diagram and achieve the same price rise for farmers, shows that the tax raises £20 whereas the deficiency payment costs £80m. This may be significant to an administration wanting to cut public expenditure. The tax reduces imports more than the deficiency payment, which may be important if the nation's balance of payments is unfavourable. But consumers are not affected directly by the deficiency payment, whereas they face higher prices and consume less when an import tax is used.

Index